DOE Simplified

Practical Tools for Effective Experimentation

DOE Simplified

Practical Tools for Effective Experimentation

Mark J. Anderson
Patrick J. Whitcomb

Productivity, Inc.
PORTLAND, OREGON

© 2000 by Productivity, Inc.

All rights reserved. No part of this book may be reproduced or utilized in any form or by any means, electronic or mechanical, including photocopying, recording, or by any information storage and retrieval system, without permission in writing from the publisher.

Most Productivity Press books are available at quantity discounts when purchased in bulk. For more information, contact our Customer Service Department (800-394-6868). Address all other inquiries to:

Productivity Press
444 Park Avenue South, Suite 604
New York, NY 10016
United States of America
Telephone: 212-686-5900
Telefax: 212-686-5411
E-mail: info@productivityinc.com
Website: www.productivitypress.com

Design and composition by William H. Brunson Typography Services
Proofreading by Susan Swanson, InPages
Printed and bound by Edwards Brothers in the United States of America

Library of Congress Cataloging-in-Publication Data

Anderson, Mark J., 1953–
DOE simplified : practical tools for effective experimentation / Mark J. Anderson, Patrick J. Whitcomb.
    p. cm.
   Includes bibliographical references and index.
   ISBN 1-56327-225-3 (alk. paper)
   1. Experimental design.   I. Whitcomb, Patrick J., 1950–.   II. Title.
QA279.A5287   2000
   658.4′033—dc21   99-086646
                                                                                     CIP

05 04 03   10 9 8 7 6 5 4

CONTENTS

Preface	ix
Introduction	xi
Flowchart Guide to *DOE Simplified*	xiii
Chapter 1: Basic Statistics for DOE	1
The "X" Factors	3
Does Normal Distribution Ring Your Bell?	5
Descriptive Statistics—Mean and Lean	7
Confidence Intervals Help You Manage Expectations	11
Graphical Tests Provide Quick Check for Normality	16
Practice Problems	20
Chapter 2: Simple Comparative Experiments	23
The F-Test Simplified	23
A Dicey Situation—Making Sure They're Fair	24
Catching Cheaters with a Simple Comparative Experiment	28
Blocking Out Known Sources of Variation	32
Practice Problems	37
Chapter 3: Two-Level Factorial Design	41
Two-Level Factorial Design—As Simple as Making Microwave Popcorn	43
How to Plot and Interpret Interactions	54
Protect Yourself with Analysis of Variance (ANOVA)	58
Modeling Your Responses with Predictive Equations	62
Diagnosing Residuals to Validate Statistical Assumptions	64
Practice Problems	68

Chapter 4: Dealing with Non-Normality via Response Transformations — 73

Skating on Thin Ice — 73
Log Transformation Saves the Data — 77
Choosing the Right Transformation — 81
Practice Problem — 84

Chapter 5: Fractional Factorials — 87

Example of Fractional Factorial at Its Finest — 88
Potential Confusion Caused by Aliasing in Lower-Resolution Factorials — 92
Plackett-Burman Designs — 98
Irregular Fractions Provide a Clearer View — 98
Practice Problem — 105

Chapter 6: Getting the Most from Minimal-Run Designs — 109

Minimal Resolution Design: The Dancing Raisin Experiment — 110
Complete Fold-Over of Resolution III Design — 115
Single Factor Fold-Over — 118
Choose a High-Resolution Design to Reduce Aliasing Problems — 119
Practice Problems — 120

Chapter 7: General Factorial Designs — 123

Putting a Spring in Your Step—A General Factorial Design on Spring Toys — 124
How to Analyze Unreplicated General Factorials — 128
Practice Problems — 132

Chapter 8: Response Surface Methods for Optimization — 135

Centerpoints Detect Curvature in Confetti — 136
Augmenting to a Central Composite Design (CCD) — 140

Chapter 9: Mixture Design — 145
Two-Component Mixture Design: Good as Gold — 146
Three-Component Design: Teeny Beany Experiment — 150

Chapter 10: Solutions to Practice Problems — 155

Chapter 11: Practice Experiments — 179
Practice Experiment #1: Breaking Paper Clips — 179
Practice Experiment #2: Hand-Eye Coordination — 181
Other Fun Ideas for Practice Experiments — 183

Appendices — 185
Appendix 1-1: Two-tailed t-Table — 185
Appendix 1-2: F-Table for 10% — 186
Appendix 1-3: F-Table for 5% — 187
Appendix 1-4: F-Table for 1% — 188
Appendix 1-5: F-Table for 0.1% — 189
Appendix 2-1: Four-Factor Screening Design — 190
Appendix 2-2: Five-Factor Screening Design — 192
Appendix 2-3: Six-Factor Screening Design — 194
Appendix 2-4: Seven-Factor Screening Design — 196

Glossary — 199
Glossary of Statistical Symbols — 199
Glossary of Terms — 200

Recommended Readings — 219
Textbooks — 219
Case-Study Articles — 219

Index — 223

About the Software — 231

PREFACE

"Without deviation from the norm, progress is not possible."
—Frank Zappa

Design of experiments (DOE) is a planned approach for determining cause and effect relationships. It can be applied to any process with measurable inputs and outputs. DOE was developed originally for agricultural purposes, but during World War II and thereafter, it became a tool for quality improvement, along with statistical process control (SPC). Until 1980, DOE was mainly used in the process industries (i.e., chemical, food, pharmaceutical), perhaps because of the ease with which engineers manipulate factors such as time, temperature, pressure, and flow rate. Then, stimulated by the tremendous success of Japanese electronics and automobiles, SPC and DOE underwent a renaissance. The advent of personal computers further catalyzed the use of these numerically-intense methods.

Our goal is to keep DOE simple and make it fun. The book is intended primarily for engineers, scientists, quality professionals, and other technical people who seek breakthroughs in product quality and process efficiency. Those of you who are industrial statisticians won't see anything new, but you may pick up ideas on translating the concepts for nonstatisticians.

By necessity, the examples in this book are generic. We believe that, without making a large stretch, you can extrapolate the basic methods to your particular application. More than two dozen case studies, covering a broad cross-section of industry, are cited in the Recommended Readings. We're certain you'll find one that you can relate to.

DOE Simplified grew from over 50 years of combined experience in providing training and computational tools to industrial experimenters. Thanks to the constructive feedback of our clients, we've made many improvements in presenting DOE since our partnership began in the mid-1970s. We've worked hard to ensure the tools are as easy to use as possible for nonstatisticians, but without compromising the integrity of the underlying principles. Our background in process development engineering helps us stay focused on the practical aspects. We've gained great benefits from formal training in statistics plus invaluable contributions from professionals in this field.

PREFACE

We are indebted to the many contributors to development of DOE methods, especially Dr. George Box and Dr. Douglas Montgomery. Most of all, we wish to acknowledge the statistical help provided by Dr. Kinley Larntz.

M.J.A. (e-mail: Mark@StatEase.com)

P.J.W. (e-mail: Pat@StatEase.com)

INTRODUCTION

"Try to make things as simple as possible, but not simpler."
—EINSTEIN

This book provides the practical tools needed for performing more effective experimentation. It examines the nuts and bolts of DOE as simply as possible—primarily by example. Our hope is that this book inspires you to master DOE!

We assume that our typical reader has little or no background in statistics. Therefore, we have kept formulas to a minimum but have used figures, charts, graphs, and checklists liberally. New terms are denoted by quotation marks and also included in a glossary for ready reference. As a spoonful of sugar to make the medicine go down, we've sprinkled the text with (mostly) relevant sidebars. Please enjoy (or forgive!) the puns, irreverent humor, and implausible anecdotes.

Furthermore, we assume that readers ultimately will rely upon software to set up experimental designs and do statistical analyses. Many general statistical packages now offer DOE for mainframe or personal computers. Other software has been developed specifically for experimenters. For your convenience, one such program is included with this book. You will find instructions for installing the software (and viewing its tutorials) at the back of the book with the CD-ROM. However, you must decide for yourself how to perform the computations for your own DOE.

We encourage readers—especially those in manufacturing areas—to master the basic quality tools of statistical process control (SPC) before taking on DOE. A good starting place is the book *SPC Simplified* (see also "Recommended Readings"). However, this is not a prerequisite, particularly for industrial researchers who primarily perform experimentation.

Chapter 1 presents the basic statistics that form the foundation for effective DOE. Readers already familiar with this material can save time by skipping ahead to Chapter 2 or Chapter 3. Others will benefit by a careful reading of Chapter 1, which begins with the most basic level of DOE: comparing two things, or two levels of one factor. You'll need this knowledge to properly analyze more complex DOEs.

Chapter 2 introduces more powerful tools for statistical analysis. You will learn how to develop experiments comparing many categories, such as various

INTRODUCTION

suppliers of a raw material. After completing this section, you will be equipped with tools that have broad application to data analysis.

Chapters 3 through 5 explain how to use the primary tool for DOE: two-level factorials. These designs are excellent for screening many factors in order to identify the vital few. They often reveal interactions that would never be found through one-factor-at-a-time methods. Furthermore, two-level factorials are incredibly efficient, producing maximum information with a minimum of runs. Most important, these designs often produce breakthrough improvements in product quality and process efficiency.

Chapter 6 introduces more complex tools for two-level factorials. Before you plow ahead, be sure to do some of the simpler factorials described in prior chapters. Practice makes perfect!

Chapter 7 goes back to the roots of DOE, which originated in agriculture. This chapter provides more general factorial tools that can accommodate any number of levels or categories. Although these designs are more flexible, they lack the simplicity of focusing on just two levels of each factor. More complications arise when you restrict the layout or order of experimentation (which is discussed in a subsection on split plots).

At this point, the book begins to push the boundaries of what can be expected from a DOE beginner. Chapters 8 and 9 definitely go beyond the boundaries of elementary tools. They offer a peek over the fence at more advanced tools for optimizing processes and mixtures. Because these last two chapters range beyond the scope of DOE "simplified," we did not include practice problems or software tools. However, advanced textbooks and computer programs are readily available to readers who want to expand their DOE horizons.

The flowchart in Figure 1 provides a chapter-by-chapter "map." At the end of Chapters 1 through 7, you will find at least one practice problem. It's definitely worth the effort to do these problems (answers to which are provided in Chapter 10). As with any new tool, the more you know about it, the more effectively you will use it. We believe that by reading this book, doing the exercises, and following up immediately with your own DOE, you will gain a working knowledge of simple comparative and factorial designs. To foster this "DOE-it-yourself" attitude, we detail several practice experiments in Chapter 11. No answers are provided, for fear that we will bias your results; but you may contact us for data from our own experiments.

FLOWCHART GUIDE TO *DOE SIMPLIFIED*

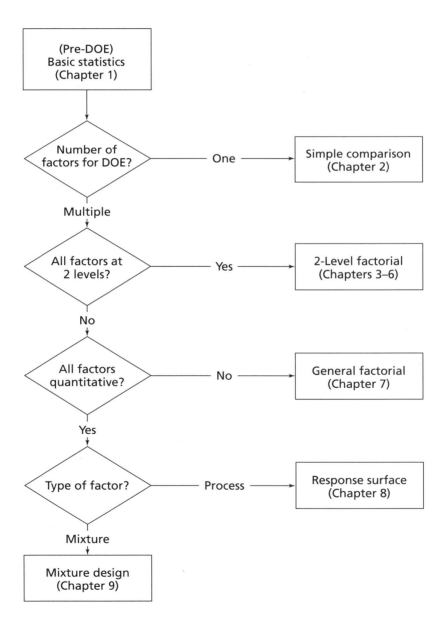

CHAPTER 1
BASIC STATISTICS FOR DOE

"One thing seems certain–that nothing certain exists."
PLINY THE ELDER, ROMAN SCHOLAR (AD 23–79)

"Statistics means never having to say you're certain."
SLOGAN ON SHIRT SOLD BY AMERICAN STATISTICAL ASSOCIATION

Most technical professionals express a mixture of fear, frustration, and annoyance when confronted with statistics. It's hard even to pronounce the word, and many people prefer to call it "sadistics"—particularly after enduring the typical college lecture on the subject. However, statistics are not evil. They are really very useful, especially for design of experiments (DOE). In this chapter we present basic statistics in a way that highlights the advantages of using them.

Statistics provide a way to extract information from data. They appear everywhere, not only in scientific papers and talks, but also in everyday news on medical advances, weather, and sports. The more you know about statistics the better, because they can be easily misused and deliberately abused.

Imagine a technical colleague calling to give you a report on an experiment. It wouldn't make sense for your colleague to read off every single measurement; instead, you'd expect a summary of the overall result. An obvious question would be how things came out on average. Then you'd probably ask about the quantity and variability of the results so you could develop some degree of confidence in the data. Assuming that the experiment has a purpose, you must ultimately decide whether to accept or reject the findings. Statistics are very helpful in cases like this—not only as a tool for summarization, but also for calculating the risks of your decision.

GO DIRECTLY TO JAIL

When making a decision about an experimental outcome, minimize two types of errors:

1. Type I ("T1"): Saying something happened when it really didn't (a false alarm). This is often referred to as the alpha (α) risk.
2. Type II ("T2"): Not discovering that something really happened (failure to alarm). This is often referred to as the beta (β) risk.

The following chart shows how you can go wrong, but it also allows for the possibility you will be correct.

Decision-Making Outcomes		What You Say Based on Experiment	
		Yes	No
The Truth	Yes	Correct	T2 Error
	No	T1 Error	Correct

The following story illustrates a Type I error. Just hope it doesn't happen to you!

> A sleepy driver pulled over to the side of the highway for a nap. A patrolman stopped and searched the vehicle. He found a powdery substance, which was thought to be an illegal drug, so he arrested the driver. The driver protested that this was a terrible mistake—that the bag contained the ashes of his cremated grandmother. Initial screening tests gave a positive outcome for a specific drug. The driver spent a month in jail before subsequent tests confirmed that the substance really was ashes and not a drug. To make matters worse, most of grandmother's ashes were consumed by the testing. The driver filed a lawsuit seeking unspecified damages. (Excerpted from a copyrighted story by the 1998 *San Antonio Express-News*.)

The "X" Factors

Let's assume you are responsible for some sort of system such as:
- Computer simulation
- Analytical instrument
- Manufacturing process
- Component in an assembled product
- Any kind of manufactured "thing" or processed "stuff"

In addition, the system could be something people-related, such as a billing process. To keep the example generic, consider the system as a black box, which will be affected by various controllable factors (see Figure 1-1). These are the inputs. They can be numerical (example: temperature) or categorical (example: raw material supplier). In any case, we'll use the letter "X" to represent the input variables.

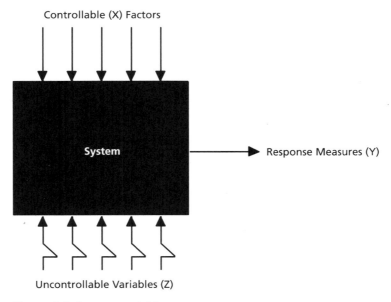

Figure 1-1. System variables

Presumably, you can measure outputs or responses in at least a semi-quantitative way. To compute statistics, you must at least establish a numerical rating, even if it's just a 1 to 5 scale. We'll use the letter "Y" as a symbol for the responses.

Unfortunately, you will always encounter variables, such as ambient temperature and humidity, that cannot be controlled or, in some cases, even identified. These uncontrolled variables are labeled "Z." They can be a major cause for variability in the responses. Other sources of variability are deviations around the set-points of the controllable factors, plus sampling and measurement error. Furthermore, the system itself may be composed of parts that also exhibit variability.

How can you deal with all this variability? Begin by simply gathering data from the system. Then make a run-chart (a plot of data versus time) so you can see how much the system performance wanders. Statistical process control (SPC) offers more sophisticated tools for assessing the natural variability of a system. However, to make systematic improvements—rather than just eliminating special causes—you must apply DOE. Table 1-1 shows how the tools of SPC and DOE differ.

Table 1-1. How DOE differs from SPC

	SPC	DOE
Who	Operator	Engineer
How	Hands-off (monitor)	Hands-on (change)
Result	Control	Breakthrough
Cause for Variability	Special (upset)	Common (systemic)

> ### TALK TO YOUR PROCESS (AND IT WILL TALK BACK TO YOU)
>
> Bill Hunter, one of the co-authors of a recommended book on DOE called *Design and Analysis of Experiments*, said that doing experiments is like talking to your process. You ask questions by making changes in inputs, then listen to the response. SPC offers tools to filter out the noise caused by variability, but it is a passive approach—used only for listening. DOE depends completely on you to ask the right questions. Asking wrong questions is sometimes called a Type III error (refer to the earlier sidebar on Type I and II errors). Therefore, subject matter knowledge is an essential prerequisite for successful application of DOE.
>
> *"When I took math class, I had no problem with the questions, it was the answers I couldn't give."*
>
> —Rodney Dangerfield

Does Normal Distribution Ring Your Bell?

When you chart data from a system, it often exhibits a bell-shaped pattern called a "normal distribution." However, not all distributions will be normal. For example, if you repeatedly roll a six-sided die, the frequency of getting 1 through 6 will be approximately equal (see bottom row of Figure 1-2). This is called a "uniform" distribution. But if you roll a pair of dice, the chances of them averaging to the extreme values of 1 or 6 are greatly reduced. The only way to hit an average of 1 from two dice is to roll two 1's (snake-eyes). On the other hand, there are three ways you can roll an average of 2: (1,3), (2,2) or (3,1). The combinations of two dice are represented by the pyramid at the top of Figure 1-2 (above the line). Average values of 1.5, 2.5, and so on now become possible. An average outcome of 3.5 is most probable from a pair of dice.

> ### DON'T SLICE, JUST DICE
>
> Rather than fight a war over a disputed island, King Olaf of Norway arranged to roll dice with his Swedish rival. The opponent rolled a double 6. "You can't win," said he. Being a stubborn Norwegian, Olaf went ahead anyway, in the hope of a tie. One die turned up 6; the other die split in two, for a total of 7 (because the opposing sides of a die always total 7.) So Norway got the island with a lucky—and seemingly impossible—score of 13. This outcome is called an "outlier," which comes from a special cause. It's not part of the normal distribution. Was it a scam? (From "The Broken Dice," by Ivar Ekeland.)

Notice how the shape of the distribution becomes more bell-shaped (normal) as you go from one die to two dice. If you roll more than two dice repeatedly, the distribution becomes even more bell-shaped and much narrower. For example, let's say you put five dice in a cup. Consider how unlikely it would be to get the extreme averages of 1 or 6—all five dice would have to come up 1 or 6 respectively. The dice play illustrates the power of averaging: the more data you collect, the more normal the distribution of averages and the closer

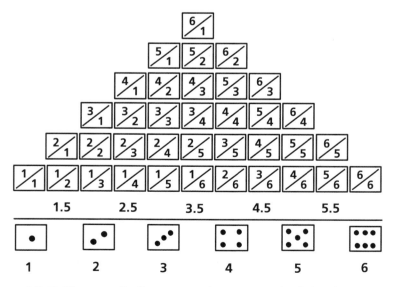

Figure 1-2. Rolling one die (bottom row) versus a pair of dice (pyramid at top)

you get to the average outcome (for the dice the average is 3.5). The distribution is "normal" because all systems are subjected to many uncontrolled variables. As in the case of rolling dice, it's very unlikely that these variables will conspire to push the response in one direction or the other. They will tend to cancel each other out and leave the system at a stable level (the mean), with some amount of consistent variability.

> ### THE ONLY THEOREM IN THIS ENTIRE BOOK
> Regardless of the shape of the original distribution of "individuals," the taking of averages results in a normal distribution. This comes from the "central limit theorem." As shown in the dice example, the theorem works imperfectly with a subgroup of two. We recommend for purposes of SPC or DOE that you base your averages on subgroups of four or more. A second aspect of the central limit theorem predicts the narrowing of the distribution (as seen in the dice example), which is a function of the increasing sample size for the subgroup. The more data you collect, the better.

Descriptive Statistics—Mean and Lean

To illustrate how to calculate descriptive statistics, let's assume your "process" is rolling a pair of dice. The output is the total number of dots that land face-up on the dice. Figure 1-3 shows a frequency diagram for 50 rolls.

Notice the bell-shaped (normal) distribution. The most frequently occurring value is 7. A very simplistic approach is to hang your hat on this outpost, called the "mode," as an indicator of the location of the distribution. A much more effective statistic for measuring location, however, is the "mean," which most people refer to as the "average." (We will use these two terms interchangeably.)

Result	Tally	Number (n)	Product
12	X	1	12
11	X	1	11
10	XXXXX	5	50
9	XXXX	4	36
8	XXXXXXXX	8	64
7	XXXXXXXXXXX	11	77
6	XXXXXXX	7	42
5	XXXXXX	6	30
4	XXXX	4	16
3	XX	2	6
2	X	1	2
Sum		50	346

Figure 1-3. Frequency distribution for 50 rolls of the dice (data from *SPC Simplified*)

EDUCATION NEEDED ON MEAN

A survey of educational departments resulted in all 50 states claiming their children to be above average in test scores for the USA. This is a common fallacy that might be called the "Lake Wobegon Effect" after the mythical town in Minnesota, where, according to author Garrison Keillor, "all the women are strong, all the men good-looking, and all the children above average."

In a related case, a company president had all his employees tested and then wanted to fire the half that were below average—believe it or not!

The formula for the mean of a response (Y) is shown below:

$$\overline{Y} = \frac{\sum_{i=1}^{n} Y_i}{n}$$

where "n" is the sample size and "i" is the individual response. The mean, or "Y-bar," is calculated by adding up the data and dividing by the number of "observations." For the dice:

$$\overline{Y} = \frac{346}{50} = 6.92$$

This is easy to do on a calculator. (Tip: if you don't have a calculator handy, look for one in the accessories of your PC's operating system. On Microsoft Windows, you'll find a calculator that computes basic statistics when you change to the scientific mode.)

STATISTICALLY (BUT NOT POLITICALLY) CORRECT QUOTES

"Even the most stupid of men, by some instinct of nature, is convinced that the more observations [n] have been made, the less danger there is of wandering from one's goal."

Jacob Bernoulli, 1654–1705

"The n's justify the means."

(Slogan on a shirt seen at an American Statistical Association meeting)

Means don't tell the whole story. For example, the authors often encounter variability in regulating room temperature in hotel meeting rooms. Typically, the temperature is frigid in the morning and steamy in the afternoon. Attendees are not satisfied that on average the temperature is about right.

The most obvious and simplest measure of variability is the "range," which is the difference between the lowest and highest response. However, this is a wasteful statistic because only two values are considered. A more efficient statistic that includes all data is "variance" (see formula below).

$$s^2 = \frac{\sum_{i=1}^{n}(Y_i - \overline{Y})^2}{n-1}$$

Variance (s^2) equals the sum of the squared deviations from the mean, divided by one less than the number of individuals. For the dice:

$$s^2 = \frac{233.68}{(50-1)} = 4.77$$

A calculator or a spreadsheet program can compute the numerator. The denominator (n – 1) is called the "degrees of freedom" (df). Consider this to be the amount of information available for the estimate of variability after calculating the mean. For example, the degrees of freedom to estimate variability from one observation would be zero. In other words, it is impossible to estimate variation. But for each observation after the first, you get one degree of freedom to estimate variance. For example, from three observations you get two degrees of freedom.

> ### A FUNCTION BY ANY OTHER NAME WILL NOT BE THE SAME
> When using statistical calculators of spreadsheet software, be careful to select the appropriate function. For example, for the numerator of the variance, you want the sum of squares (SS) *corrected for the mean*. For example, Microsoft Excel 2000 offers a SUMSQ worksheet function, but this does not correct for the mean. The proper function is DEVSQ, which does correct for the mean.

Variance is the primary statistic used to measure variability, or dispersion, of the distribution. However, to get units back to their original (not squared) metric, it's common to report the "standard deviation" (s). This is just the square root of variance:

$$s = \sqrt{\frac{\sum_{i=1}^{n}(Y_i - \overline{Y})^2}{n-1}}$$

For the dice:

$$s = \sqrt{4.77} = 2.18$$

> ### SAMPLING—A MEATY SUBJECT
>
> The symbol "s" is used when we calculate standard deviation from sample data. Some statistical calculators offer this function under the label of "σ_{n-1}," where the Greek letter sigma refers to true value of standard deviations for the "population." Sampling is a necessity in process improvement, because the population would be all the possible outcomes, which are unknown and unknowable. Even if you could test the entire population, it wouldn't be efficient to do so. This became a huge political issue for the year 2000 US census takers who wanted to use sampling rather than counting every individual in the entire population. The statistical treatment for populations is simpler for sample data, but because it's not relevant for industrial experimenters, we won't spend time on the population statistics. You can get this information from any book on basic statistics.
>
> *"You do not have to eat the whole ox to know the meat is tough!"*
> —Shakespeare

Confidence Intervals Help You Manage Expectations

In the dice example, the sample mean was calculated from 50 rolls. What happens if the pair of dice is rolled another 50 times? What if you repeat this experiment many times? You're encouraged to try this, but you can probably guess the outcome—the means will vary somewhat around the true value of 7 (assuming perfect dice). This variation, measured by taking a standard deviation of the means, is called the "standard error" (SE) of the mean.

It would be a terrible waste of time to repeat an experiment many times to get an estimate of the standard error. To save time, statisticians use an approximation based on part two of the central limit theorem (see sidebar):

$$SE \equiv s_{\bar{y}} \cong \sqrt{\frac{s^2}{n}}$$

For the dice:

$$SE \cong \sqrt{\frac{4.77}{50}} = 0.31$$

The standard error gets smaller as the number sampled (n) gets larger. You get what you pay for; the more you sample, the more precisely you can estimate the 'true' outcome.

Very few people understand standard error, but almost everybody recognizes a related statistic called the "confidence interval," particularly in election years. Any political poll that calls itself 'scientific' reveals its so-called margin of error. For example, a pollster may say that the leading candidate got a 60% approval rating, plus or minus 3%. In most cases, when such a margin or interval is reported, it's based on a confidence level of 95%. Intervals constructed in this way will contain the true value of the calculated statistic 95% of the time.

The formula for the confidence interval (CI) of a mean is:

$$CI = \overline{Y} \pm t \times SE$$

where "t" is a factor that depends on the confidence desired and the degrees of freedom generated for the estimate of error. It represents the number of standard deviations from the estimated mean. The t-statistic can be looked up in the table provided in the Appendix 1-1.

> ### *BELIEVE IT OR NOT: A UNIVERSITY STUDENT WHO'S AN EXPERT ON BEER*
>
> An Oxford graduate in chemistry and mathematics who worked at Guinness brewery in Dublin, Ireland, characterized the t-distribution in 1908. His name was W.S. Gosset, but out of modesty, or perhaps due to restrictions from Guinness, he published under the pen-name 'Student.' Prior to Gosset, statisticians directed their attention to large samples of data, often with thousands of individuals. Their aims were primarily academic. Guinness couldn't afford to do pure research. They did only limited field trials on natural raw materials such as barley. Gosset/Student discovered that
>
> "As we decrease the number of experiments, the value of the standard deviation found from the sample of experiments becomes itself subject to increasing error."
>
> He developed practical tools for dealing with variability in soil and growing conditions so the decisions could be made with some degree of confidence. Here's how Gosset/Student described the distribution (t) of small samples:
>
>> "If the sample is very small, the 'tails' of its distribution take longer to approach the horizontal and its 'head' is below the head of the normal distribution; as the number in the sample increases, the tails move 'inwards' towards the tails of the normal distribution and the 'head' of the curve moves toward the 'head' of the normal distribution."
>
> (From *Invitation to Statistics*, Gavin Kennedy, Martin Robinson & Co., Oxford, 1983)

As you can see in Figure 1-4, the t-distribution looks like a normal curve with 'heavy' tails (fatter, longer). However, for practical purposes, when you collect a large amount of data, say n > 30, the t-distribution becomes normal. Figure 1-5 shows the percent of individual outcomes as a function of standard deviation for the normal curve (or for the t-distribution based on a very large sample size).

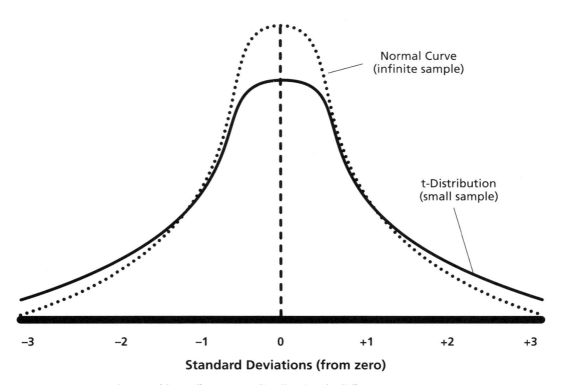

Figure 1-4. Normal curve (dotted) versus t-distribution (solid)

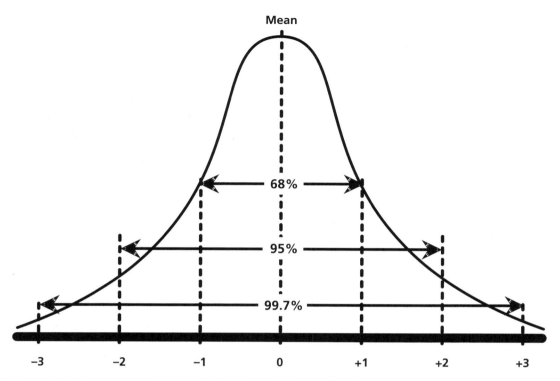

Figure 1-5. Percentage of individuals falling within specified limits of standard deviation for normal distribution (or t-distribution based on large sample size)

Figure 1-5 demonstrates that the majority of individuals (specifically 68%) from a given population will fall within one standard deviation of the mean (the zero point). By expanding the range of standard deviation to plus or minus two, we can include 95% of the population. Any individuals outside of this wider range certainly could be considered to be special. Therefore, for large samples, it makes sense to set the critical value of t at 2 when constructing a confidence interval. In the example of the dice:

$$CI = 6.92 \pm 2(0.31) = 6.92 \pm 0.62$$

Intervals constructed in this manner will include the true population mean 95% of the time. The calculated range, from 6.30 (6.92 minus 0.62) to 7.54 (6.92 plus 0.62)—commonly called a 95% confidence interval—does encompass the theoretical outcome of 7. (Thank goodness!)

Confidence intervals are extremely useful for experimenters, because they allow an assessment of uncertainty. Using statistics allows you to quantify probabilities for error and play a game of calculated risks.

> ### HOW NOT TO INSPIRE CONFIDENCE
> A young engineer, obviously not well-schooled in statistics, made this statement about a problem encountered while testing a new piece of equipment: "This almost always hardly ever happens."
>
> What do you suppose is the probability of "this" happening again?

Graphical Tests Provide Quick Check for Normality

The basic statistical tools discussed in this section are very powerful. However, they all depend on the assumption of normality. As anyone who does technical work knows, it always pays to check assumptions before making a final report. So before going any further, it will be worthwhile to learn how to check your data with a special type of graph called a "normal plot." This graph is really handy, because normally distributed data will fall nicely in line when plotted properly. You will see many a normal plot—and its cousin the "half-normal" plot—throughout this book.

The normal plot requires a properly scaled template called "probability paper." Long ago you would get this from a commercial vendor of graph paper, but now most people generate normal plots from statistical software. However, it's not that hard to make it yourself. The x-axis is easy—it's made just like any other graph with regularly spaced tic marks. The y-axis, on the other hand, is quite different, because it's scaled by "cumulative probability." This is a measure of the percentage of individuals expected to fall below a given level, benchmarked in terms of standard deviation. For example, as shown by the shaded area in Figure 1-6, the cumulative probability at the benchmark of 1 standard deviation is approximately 84%.

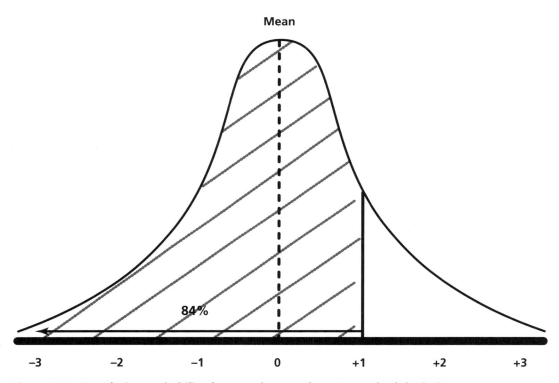

Figure 1-6. Cumulative probability (area under curve) at +1 standard deviation

The cumulative probability can be determined by measuring the proportion of area under the normal curve that falls below any given level. Statisticians have enumerated this in great detail. Table 1-2 provides the highlights.

Table 1-2. Cumulative probability versus number of standard deviations (from the mean)

Standard Deviations	Cumulative Probability
−2.0	2.3%
−1.0	15.9%
0.0	50.0%
1.0	84.1%
2.0	97.7%

We are now ready to tackle the mysterious y-axis for probability paper. Figure 1-7 shows how this can be done by putting standard deviations on a linear scale at the right side and recording the associated cumulative probability on the left side. We added a lot more detail to make the plot more usable. As you can see, the cumulative probability axis is very nonlinear.

Now that we've done all this work, let's go for the payoff: checking data for normality. The following 10 weights (in pounds) come from a random sample of men at the 25th reunion of an all-boys high school class: 199, 188, 194, 206, 235, 219, 200, 234, 171, 172. Various statistics could be computed on these weights (and snide comments made about their magnitude), but all we want to do is check the data for normality. A few things must be done before plotting the data on the probability paper:

1. Sort the n datapoints in ascending order.
2. Divide the 0 to 100% cumulative probability scale into n segments.
3. Plot the data at the midpoint of each probability segment.

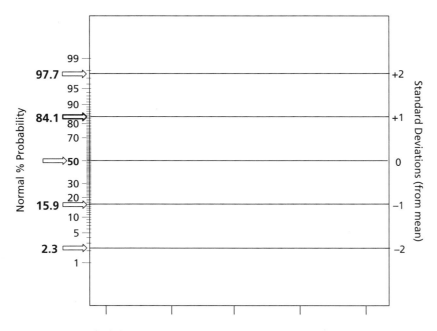

Figure 1-7. Probability paper with standard deviation scale at the right

Table 1-3. Values to plot on probability paper

Point	Weight	Cumulative Probability
1	171	5%
2	172	15%
3	188	25%
4	194	35%
5	199	45%
6	200	55%
7	206	65%
8	219	75%
9	234	85%
10	235	95%

In this case, the sample size (n) is 10, so each probability segment will be 10% (100/10). The lowest weight will be plotted at 5%, which is the midpoint of the first segment. Table 1-3 shows this combination and all the remaining ones.

Now all we need do is plot the weights on the x-axis of the probability paper and the cumulative probabilities on the y-axis (see Figure 1-8).

The interpretation of the normal plot is somewhat subjective. Look for gross deviations from the linear, such as a big "S" shape. A simple way to check for linearity is to apply the "pencil test." If you can cover all the points with a pencil, the data are normal. It's OK if only a portion of some points touch the pencil. Don't be too critical! According to this criteria, the weight data passes the pencil test. Therefore, the alumni apparently conform to the normal bell-shaped curve weight-wise (and for the heavier men, shape-wise!).

This concludes our coverage of the basic statistical tools that form the foundation of DOE. You are now ready to move on to DOE procedures for making simple comparisons, such as which material works best or whether you will benefit by changing suppliers.

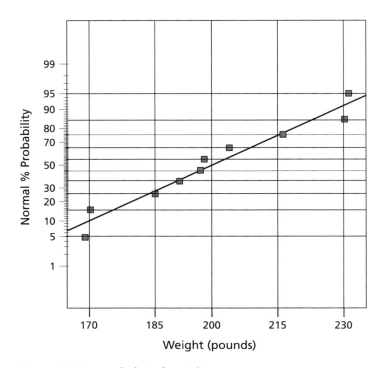

Figure 1-8. Normal plot of weights

Practice Problems

To practice using the statistical techniques you learned in Chapter 1, work through the following problems.

PROBLEM 1-1

You are asked to characterize the performance of a motor-shaft supplier. The data shown below is a measure of the endplay:

61, 61, 57, 56, 60, 52, 62, 59, 62, 67, 55, 56, 52, 60, 59, 59, 60, 59, 49, 42, 55, 67, 53, 66, 60.

Determine the mean, standard error, and approximate 95% confidence interval for the endplay using the guidelines given in this chapter. (Suggestion: enter the data in a handheld or software calculator that can compute standard deviation, or use spreadsheet software with similar capabilities.)

PROBLEM 1-2

The alumni from the case described at the end of the chapter reminisced about their weights at the time of graduation. One of them contacted the school nurse and dug up the old records. These are shown below:

153, 147, 148, 161, 190, 167, 155, 178, 130, 139.

Simple inspection of these results versus those reported earlier will reveal an obvious impact of aging, so we needn't bother doing any statistics (enough said!). Your job is to construct a normal plot of these 10 datapoints. (Suggestion: to save time, re-use the normal plot shown earlier in Figure 1-8. Just subtract 40 from each tic mark. Then plot the new points with a red pen or pencil so you can see them clearly.) Do you see anything grossly abnormal about the data?

THE "BODY" IMPOLITIC

Former wrestler, Governor Jesse Ventura, of the author's home state of Minnesota, described an unfavorable political poll as "skewered." The proper term for a non-normal distribution that leans to the left or right is "skewed." However, Governor Ventura, who tries to stay in the center, may be on to something with his terminology.

"Keep it simple and stupid."
—Jesse Ventura

CHAPTER 2
SIMPLE COMPARATIVE EXPERIMENTS

"Many of the most useful designs are extremely simple."
SIR RONALD FISHER

We will now look at a method for making simple comparisons of two or more "treatments." It's called the "F-test," after Sir Ronald Fisher, a geneticist who developed the technique for application to agricultural experiments. The F-test compares the variance among the treatment means versus the variance of individuals within the specific treatments. High values of F indicate that one or more of the means differ from another. This can be very valuable information when, for example, you must select from several suppliers, or materials, or levels of a process factor. The F-test is a vital tool for any kind of DOE, not just simple comparisons, so it's important to understand it as fully as possible.

PUT ON YOUR KNEE-LENGTH BOOTS!

In his landmark paper on DOE entitled "The Differential Effect of Manures on Potatoes," Fisher analyzes the impact of varying types of animal waste on yield of spuds. The agricultural heritage explains some of the farm jargon you see in writings about DOE: blocks, plots, treatments, environmental factors, etc. It's a lot to wade through for nonstatisticians, but worth the effort.

The F-Test Simplified

Without getting into all the details, the following formula for F can be derived from part two of the central limit theorem:

$$F \equiv \frac{n s_{\bar{y}}^2}{s_{pooled}^2}$$

This formula assumes that all samples are of equal size n. You might think of F as a ratio of signal (differences caused by the treatments) versus noise. The F-ratio increases as the treatment differences become larger. It becomes more sensitive to a given treatment difference as the sample size (n) increases. Thus, if something really does differ, you will eventually find it if you collect more data. On the other hand, the F-ratio decreases as variation (s^2_{pooled}) increases. This noise is your enemy. Before you even begin your experimentation, do what you can to dampen system variability via statistical process control (SPC), quality assurance on your response measurement, and control of environmental factors.

If the treatments have no effect, then the F-ratio will be near a value of 1. As the F-ratio increases, it becomes less and less likely that this could occur by chance. With the use of statistical tables such as those provided in the Appendix, you can quantify this probability ("p"). The "p-value" becomes a good "bottom-line" indicator of significance. When the F-ratio gets so high that the p-value falls below 0.05, then you can say with 95% confidence that one or more of the treatments is having an effect on the measured response. This still leaves a 5% "risk" that noise is the culprit. If 5% risk is too much, you can set a standard of 1%, thus ensuring 99% confidence. Conversely, you may want to live dangerously by taking a 10% risk (90% confidence). It's your choice. Our choice for examples shown in this book will be 95% confidence for all tests and intervals.

> ### *ANOVA: NEITHER A CAR, NOR AN EXPLODING STAR*
> The F-test uses variance as its underlying statistic. Therefore, statisticians call the overall procedure "analysis of variance," or ANOVA. When this is applied to simple comparisons, it's called a "one-way" ANOVA. With the advent of built-in spreadsheet functions and dedicated statistical software, ANOVA can be accomplished very easily. However, it still can be somewhat intimidating for nonstatisticians.

A Dicey Situation—Making Sure They're Fair

Let's look at an example: further experimentation with dice. One of the authors purchased a game called "Stack®" by Strunk Games. The game comes with 56 six-sided dice in four colors (14 each).

SIMPLE COMPARATIVE EXPERIMENTS

Table 2-1. Frequency distribution for 56 rolls of the dice

Result	White (1)	Blue (2)	Green (3)	Purple (4)
6	6/6	6/6	6/6	6
5	5	5	5	5
4	4	4/4	4/4	4
3	3/3/3/3/3	3/3/3	3/3/3/3	3/3/3/3/3
2	2/2/2	2/2/2/2	2/2/2/2	2/2/2/2/2
1	1/1	1	1	1
Mean ($\bar{Y}$)	3.14	3.29	3.29	2.93
Var. (s^2)	2.59	2.37	2.37	1.76

Let's assume that it's advantageous to get dice that tend to roll high. The first question is whether the change in colors causes a significant effect on the average outcome. Table 2-1 shows the results obtained by dumping the dice on a desk. The bag was first shaken to thoroughly mix the dice, thus assuring a "random" toss.

Notice that the data itself forms a histogram when viewed sideways. You can see the relative uniform shape (non-normal) of the distribution for individual dice. Furthermore, you will observe a great deal of overlap between colors and tosses. It's hard to see any differences, but why take a chance? It pays to perform ANOVA.

DON'T GET PARANOID ABOUT PATTERNS IN DATA

There appears to be an inordinate number of 3's and 2's in Table 2-1, but this was just a chance occurrence. If you don't believe this, see Problem 2-2 at the end of this chapter, which shows results from a second toss. Coincidences occur more often than most people think. It's best to be skeptical until you apply statistical tools such as ANOVA.

"You can see a lot just by looking."

—Yogi Berra

Let's get started on the calculation of F. The simplest part is determining the sample size (n). The answer is 14—the number of dice of each color. Next, the variance (s^2) of the means must be calculated. This can be done on any statistical calculator, standard spreadsheet program, or by chugging through the following equation.

$$S_{\bar{y}}^2 = \frac{(3.14 - 3.1625)^2 + (3.29 - 3.1625)^2 + (3.29 - 3.1625)^2 + (2.93 - 3.1625)^2}{4 - 1} = 0.029$$

where the value 3.1625 is the grand mean of all the responses, or equivalently, the average of the 4 group means (3.14, 3.29, 3.29, 2.93). (The extra digits on the grand mean are carried along for the sake of accuracy.) Finally, the last element in the F equation, the variance of the individuals pooled (S^2_{pooled}), is computed by simply taking the mean of the variance within each type.

$$S^2_{pooled} = \frac{2.59 + 2.37 + 2.37 + 1.76}{4} = 2.28$$

Therefore, for toss one of the colored dice:

$$F = \frac{n * S_{\bar{y}}^2}{S^2_{pooled}} = \frac{14 * 0.029}{2.28} = 0.18$$

The value of 14 for n (sample size) represents the number of dice of each color. The resulting F-ratio falls below 1, which implies that the variation between mean results by color is not excessive when compared with the variation within each color. We provide F-tables in the Appendix, but it's not worth looking at them at this point because the F is below a value of 1 (we will get back to the tables in Part 2 of this example). It's correct to assume that the results are not significant. Therefore, it appears safe to go ahead and play the game.

NEVER SAY NEVER, BUT IT'S OK TO STATE A DEFINITE MAYBE

Statisticians think of experimentation as a "hypothesis test." Their approach is to assume that there's no effect. This is the "null" hypothesis. It must be disproved beyond a specified risk, such as 5%. This can be likened to the jury system of law, where the standard is: innocent until proven guilty beyond a shadow of a doubt. A subtlety of this statistical approach is that you can never prove the null—no effect. It would require an infinite sample size to do so. For example, in the dice case, a very subtle doctoring could be obscured by natural variability over many cycles of testing, but in the long run it could provide enough of an advantage to make a good living for a crooked gambler.

SIMPLE COMPARATIVE EXPERIMENTS

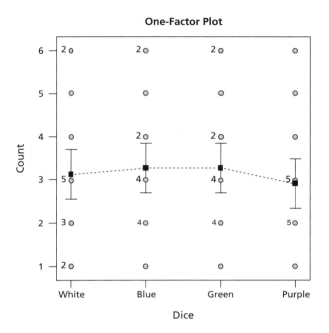

Figure 2-1. Effects plot for dice of various colors (for fits toss only)

Figure 2-1 shows graphs of the data with "least significant difference" (LSD) bars, a form of confidence interval, superimposed. These are set at 95% confidence. Don't be alarmed that results fall outside the interval because it's related to the average outcome, not the individuals. The more data you collect, the narrower the interval becomes.

The number next to some points indicates repeated results; for example, two 6's, five 3's, three 2's, and two 1's turned up for the white dice 1 on the first toss. The middle line on the bars represents the mean value for the dice subgroup. The bars are set to provide 95% confidence. They obviously overlap, thus confirming the lack of significance in the outcome of the colored dice experiment. However, in this case, you already know this from the insignificant F-test.

The formula for the LSD, with equal sample sizes (n), is:

$$\text{LSD} = t_{critical} \times s_{pooled}\sqrt{2/n}$$

This formula is derived from the equation for a confidence interval shown in Chapter 1. As discussed earlier, "t" is a factor that depends on the confidence desired and the degrees of freedom generated for the estimate of standard deviation, pooled from all the subgroups.

> ### BE CAREFUL WHEN USING LSD!
>
> Using LSD without the protection of a significant F-test could be a real "downer" because of the risk of getting false positive differences. For example, if you were to compare 10 things two at a time, you'd get 45 combinations. Chances are very good that by arbitrarily focusing only on the highest and lowest outcomes you would get a significant test even from a random set of data. So "just say no" to pair-wise testing before conducting the F test. (There are a number of special procedures for multiple pair-wise testing in the reference by Box in the Recommended Readings section of this book.)
>
> You may want to listen to "Lucy in the Sky with Diamonds" by the Beatles at this point!

A sample calculation for LSD is provided in Part 2 of this example, which illustrates what happens when you get a significant F-test. Read on for a thrilling conclusion to this case study!

Catching Cheaters with a Simple Comparative Experiment

To add some spice to this discussion, let's assume that a nefarious gamer messes with the dice in an attempt to gain an advantage. This could be done in any number of ways: by doctoring the surfaces, rounding the edges, or "loading" the dice with tiny weights. The actual cheating method will not be revealed yet; just the results seen in Table 2-2. The results from just one toss were sufficient to catch the cheaters.

SIMPLE COMPARATIVE EXPERIMENTS

> ### THE ONE SURE THING ABOUT GAMBLING: SOMEONE WILL TRY TO CHEAT
>
> Archeologists discovered several pairs of sandstone dice in an ancient Egyptian tomb. These dice, now residing in a Chicago museum, are weighted to favor twos and fives. It seems that loaded dice are as old as the pyramids. The statisticians have a term that relates to things like this that don't come out true—it's called "bias." It's safe to say that the objective is to eliminate bias if at all possible.
>
> (From "The Roll of the Dice," March '98, *Technology Magazine*)
>
> *"If the dice roll long enough, the man who knows what numbers are favored is going to finish with a fatter bankroll than when he started."*
>
> Gambling authority John Scarne

The equation expressed in terms of statistical functions is:

$$F = \frac{ns_{\bar{Y}}^2}{s_{pooled}^2} = \frac{14 * Var(2.50, 3.93, 4.93, 2.86)}{Mean(2.42, 2.38, 2.07, 3.21)} = \frac{14 * 1.21}{2.52} = 6.71$$

A reminder: the value of 14 for n (sample size) represents the number of dice of each color.

F-tables for various levels of risk are provided in the Appendix. You can use these to determine whether your results are significant. But you must first determine how much information, or degrees of freedom (df), went into the

Table 2-2. Frequency distribution for 56 rolls of the doctored (?) dice

Result	White (1)	Blue (2)	Green (3)	Purple (4)
6	6	6/6/6	6/6/6/6/6	6
5	5	5/5	5/5/5/5	5/5
4	4	4/4/4	4	4/4/4
3	3/3	3/3/3/3	3	3
2	2/2/2/2/2	2		2/2
1	1/1/1/1	1	1	1/1/1/1
Mean ($\bar{Y}$)	2.50	3.93	4.93	2.86
Var. (s^2)	2.42	2.38	2.07	3.21

calculations for the variances. Calculation of the df is not complicated but it can be tedious. See the sidebar if you want the details for this case, but the appropriate df's are 3 (4 − 1) and 52 (4 times 14 − 1) for the variances between treatments (numerator of F) and within treatments (denominator), respectively. If you go to column 3 and row 40 and row 60 (there is no row for 52) for the 5% table, you will see F-values of 2.839 and 2.758, respectively. To be conservative, let's use the value of 2.839. Because this critical value is exceeded by the actual F of 6.71, you can be more than 95% confident of significance for the test. Just out of curiosity, look up the critical F for 1% risk. You should find a value of about 4.2, which is still less than the actual F, thus indicating significance at greater than 99% confidence. It really looks bad for the cheaters!

> ### DEGREES OF FREEDOM—WHO NEEDS 'EM?
> ### (WARNING: STATISTICAL DETAILS FOLLOW)
>
> In the two dice experiments, the variance for the treatments (numerator of F) is based on four numbers (means for each color of dice), so just subtract 1 to get the df of 3. The pooled variance within each treatment (denominator of F) is accumulated from 4 groups of 14 results (56 total), but 4 df must be subtracted for calculating the means, leaving 52 df for estimating error.

With the significant F-test as protection, we now use LSD as a comparative tool:

$$LSD = t_{critical} s_{pooled} \sqrt{2/n}$$

where the pooled standard deviation (spooled) is calculated by taking the square root of the average variance:

$$s_{pooled} = \sqrt{(2.42 + 2.38 + 2.07 + 3.21)/4} = \sqrt{2.52} = 1.59$$

Recall that this estimate of error is based on 52 "degrees of freedom" (df). As a rough rule of thumb, when you have more than 30 df, which we easily exceed, the t-distribution becomes approximately normal. In Figure 1-5 (see p. 15), you can see that about 95% of a population falls within two standard deviations.

SIMPLE COMPARATIVE EXPERIMENTS

Thus, a reasonable approximation for the critical t at 95% confidence will be 2. Plugging in the subgroup size (n) of 14, we get:

$$\text{LSD} \cong 2*1.59\sqrt{2/14} = 1.20$$

The LSD of 1.2 is exceeded by the 1.43 difference between white and blue (3.93 − 2.50). Therefore this comparison is statistically significant. You can see this in Figure 2-2: The LSD bars for the white and blue groups of dice do not overlap. The graph reveals that the blue and green dice are rolling much higher than the white and purple. Further tosses presumably would strengthen this finding, but it wasn't deemed necessary.

In this case, two cheaters (playing the blue and green dice) conspired against the other two (white and purple). While one partner in crime distracted the victims, the other turned most of the white and purple dice down one dot and most of the blue and green up one dot. To allay suspicion, they did not

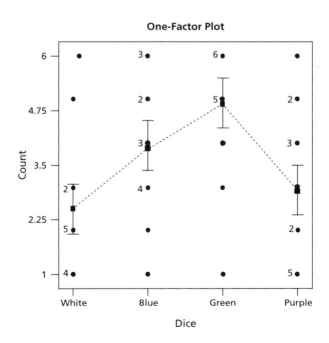

Figure 2-2. Effects plot for unfair dice game

change a die that was the last of its kind; for example, the last 3. This all had to be done in a hurry, so mistakes were made, which contributed to the overall variability in response. Despite this, the scheme proved effective, perhaps too much so, because all but young children would see through it fairly quickly. Professional cheaters would be much more subtle. Nevertheless, by accumulating enough data, statistical analysis would reveal the effect.

Blocking Out Known Sources of Variation

Known sources of variation caused by changes in personnel, materials, or machinery can be a real nuisance. Fortunately for us, statisticians such as Fisher developed techniques to "block" out these nuisance variables. In a landmark field trial on barley (for making beer!) in the authors' home state of Minnesota, agronomists grew 10 varieties of the crop at 6 sites in the early 1930s. Within each block of land, the barley was planted at random locations. The yields varied considerably due to variations in soil and climate, but the ranking of barley types remained remarkably consistent. This is called a "randomized block" experiment. A good rule for DOE is to block what you can and randomize what you cannot. There are many ways to incorporate blocking in a DOE, but we will illustrate only the more common and simple approaches.

Blocking is an especially effective tool for experiments that involve people. Each person will behave differently, but in many cases a consistent pattern will emerge. Box (see Recommended Readings) describes a clever approach for blocking out variation caused by the natural differences between boys. The objective is to measure wear of two alternative raw materials for shoe soles. The problem is that boys vary tremendously in terms of energy, from hyperactives at one end of the normal curve to TV-viewing "couch potatoes" at the other extreme. The resulting variation in shoe wear caused by boy-to-boy difference was blocked by giving each boy one shoe of each material, applied on a random basis to the left or right shoe. A clear difference in materials emerged despite an overwhelming variation from one boy to the next.

The following example illustrates the procedure for blocking. It involves the same dice used in previous studies, but with different rules. A group of preschool children were asked to pick out and stack three dice with 1, 2, 3, 4, 5, or 6 dots (any color). Because young children develop at such differing rates, the time needed to accomplish this task varied widely. However, by repeating the entire exercise for each child, the differences between children were blocked out. The final results were somewhat surprising.

> ### LEARNING FROM SIMPLE COMPARATIVE EXPERIMENTATION
>
> The famous child psychologist Jean Piaget suggested that thinking is a flexible, trial-and-error process. He observed that children, at their earliest stage of development, experiment with objects and learn from experience. For example, they often find it difficult at first to grasp the concept of volume. To confirm this observation, preschoolers were shown a tall, narrow container filled with water. They were asked to guess what would happen when the teacher poured the water into a broad and shallow bowl of equal volume. Here are the more interesting hypotheses:
>
> "It will be a volcano—it will start on fire."
> "It won't fit. It will explode all over."
> "It will be juice."
> "It will turn pink."
> "It will bubble."
> "It will turn red. It got full! It's magic. The teacher's magic!"
>
> Only a few of the children guessed correctly that water would not overflow. Presumably many of the others learned from their mistakes, but for some of the preschoolers it may take a few repetitions and further cognitive development for the concept of volume to completely sink in.
>
> *"When consequences and data fail to agree, the discrepancy can lead, by a process called induction, to modification of the hypothesis."*
>
> —Box, Hunter, and Hunter

Table 2-3 shows the data for four children. You can assume that each result is actually an average from several trials, because individual times would normally vary more than those shown. The last row shows how each child differs from the overall average of 36.9 seconds. For example, the K1 child completed the tasks 18.47 seconds faster on average than the group as a whole. These differences are substantial relative to the differences due to the various dot patterns. For example, child K3 evidently required some coaxing to do the task at hand, because the times far exceed those of the other individuals.

Table 2-3. Times (in seconds) for stacking three dice with same number of dots

Child (block)	K1	K2	K3	K4	Within-Dot Mean	Within-Dot Variance
1 dot	7.2	13.2	39.9	22.2	20.6	203.3
2 dots	10.0	21.6	45.3	24.1	25.3	216.8
3 dots	25.6	36.2	79.7	44.6	46.5	548.3
4 dots	15.2	30.0	54.5	32.9	33.1	262.5
5 dots	33.0	48.1	90.8	52.7	56.1	604.7
6 dots	19.5	32.0	65.0	42.0	39.6	370.9
Mean	18.4	30.2	62.5	36.4	36.9	367.8
Difference from overall mean	−18.47	−6.71	25.64	−0.46	0.0	

The F-test with no correction for blocks is shown below:

$$F = \frac{ns_{\bar{y}}^2}{s_{pooled}^2} = \frac{4*\text{Var}(20.6, 25.3, 46.5, 33.1, 56.1, 39.6)}{367.8} = \frac{4*176.9}{367.8} = 1.92$$

This F-value is based on 5 df for the between means comparison (numerator) and 18 df for the within means comparison. The critical F for 5% risk is 2.773, so the actual F fails to reject the null hypothesis at the 95% confidence level. In other words the results are not significant. Figure 2-3 reveals a pattern, but it's obscured statistically by the variability between children. How can we properly recognize the pattern statistically?

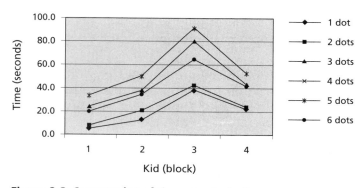

Figure 2-3. Scatter-plot of times to stack dice

Table 2-4. Times (in seconds) for stacking 3 dice (corrected for blocks)

Child (block)	K1	K2	K3	K4	Within-Dot Mean	Within-Dot Variance
1 dot	25.7	19.9	14.3	22.7	20.6	23.6
2 dots	28.5	28.3	19.7	24.6	25.3	17.0
3 dots	44.1	42.9	54.0	45.1	46.5	25.7
4 dots	33.7	36.7	28.8	33.4	33.1	10.5
5 dots	51.5	54.8	65.2	53.2	56.1	38.0
6 dots	38.0	38.7	39.4	42.5	39.6	3.9
Mean	36.9	36.9	36.9	36.9	36.9	23.7
Difference from overall mean	0.00	0.00	0.00	0.00		

The answer is to remove the variability between children by subtracting the difference shown on the last line of Table 2-3 from the raw data in the associated columns. This is how you can block out a known source of variation. The results are given in Table 2-4.

The means for each dot pattern (1 through 6) remain the same, but notice the huge reduction in within-dot variance after removal of blocks. It drops more than tenfold, from 367.8 down to 23.7. Obviously this will have a very beneficial impact on the F-ratio. The calculation is:

$$F = \frac{4*176.9}{23.7} = 29.8$$

Notice that the numerator remains unchanged from the unblocked case, because the treatment means are unaffected. The degrees of freedom for the numerator also remain the same as before, at 5 df. However, the degrees of freedom for the denominator of F must be reduced because we took the block means into account. Specifically, the calculation of the 4 block means causes a loss of 3 df (n − 1 with n = 4) for calculating error, so only 15 df remain. This loss of information for estimating error must be accounted for when looking up the critical value for F. On the 5% table, you will find a value of 2.901 under column 5 and row 15. The actual F of 29.8 far exceeds this critical F, so the outcome is statistically significant. In this case, blocking proved to be the key to success.

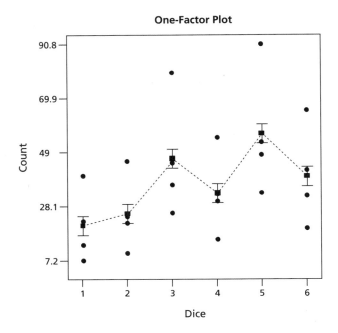

Figure 2-4. Effects plot of times to stack dice (corrected for blocks)

Figure 2-4 shows the times as a function of the number of dots. It does not show the blocks, but they are accounted for in calculating the LSD bars. Aided by this statistical tool, you can see the unexpected outcome: the children found it most difficult to pick out the 5-spot dice. They were very quick to identify the 1-spot dice, significantly so in relation to the 5 (also the 3, 4, and 6). The 3-spot seemed to cause more trouble than the 4-spot dice.

Many more children would need to be tested to confirm these findings. The point of this exercise is simply to show the advantage of blocking out known sources of variation. Blocking can be applied to any kind of DOE, not just simple comparisons such as those illustrated in this chapter. It often reveals results that would otherwise be obscured.

SIMPLE COMPARATIVE EXPERIMENTS

Practice Problems

PROBLEM 2-1

Three bowlers must compete for the last position on the company team. They each bowl 6 games (see data in Table 2-5). The best bowler will fill the opening, with the runner-up as the alternate. The worst bowler is out.

Assume that you are the captain. You know better than to simply pick the bowler with the highest average score and drop the person who scores lowest. Maybe it's a fluke that Mark scored highest and Pat's score is low. Are the scores significantly different, given the variability in individual scores? Do an analysis of variance. If it produces significant results, the choice may be clear; otherwise, the bowlers must return to the alley for more games. (Suggestion: use the software provided with the book. First do the one-factor tutorial that comes with the program. It's keyed to the data in Table 2-5. See the accompanying instructions for software installation and details on the tutorial files.)

Table 2-5. Bowling scores

Game	Pat	Mark	Shari
1	160	165	166
2	150	180	158
3	140	170	145
4	167	185	161
5	157	195	151
6	148	175	156
Mean	153.7	178.3	156.2

Table 2-6. Frequency distribution for 50 rolls of the dice, second toss

Result	White (1)	Blue (2)	Green (3)	Purple (4)
6	6/6	6/6/6	6/6/6	6/6/6
5	5/5/5/5	5	5/5/5	5/5/5
4	4/4/4	4/4	4/4	4/4/4
3	3/3/3	3/3/3/3	3/3	3
2	2/2	2/2/2	2/2	2
1		1	1/1	1/1/1
Mean ($\bar{Y}$)				
Var. (s^2)				

PROBLEM 2-2

To double-check the dice comparison detailed in Table 2-1, the players tossed the dice a second time. The data is shown in Table 2-6.

Do an analysis of variance for the second toss according to the procedures shown earlier. Do you have any reason to dispute the earlier conclusion that color of dice does not significantly affect the outcome (so long as no one cheats)? (Suggestion: follow up by using the software provided with the book. Set up a one-factor design similar to that shown in the tutorial that comes with the program. After doing the ANOVA, generate an effects plot with the LSD bars.)

PROBLEM 2-3

In problem 1-1 you analyzed performance of a supplier. Let's assume that you get three more incoming shipments. The delivery person observes that something looks different than before. You decide to investigate using analysis of variance. The data are collected in Table 2-7.

Do you see a significant difference between lots? If so, which ones stand out, either higher or lower than the others? (Suggestion: use the software provided with the book. Set up a one-factor design similar to that shown in the tutorial that comes with the program. After doing the ANOVA, generate an effects plot with the LSD bars.)

SIMPLE COMPARATIVE EXPERIMENTS

Table 2-7. Data from four incoming shipments

Lot	Data
A	61,61,57,56,60,52,62,59,62,67,55,56,52,60,59,59,60,59,49,42,55,67,53,66,60
E	56,56,61,67,58,63,56,60,55,46,62,65,63,59,60,60,59,60,65,65,62,51,62,52,58
I	62,62,72,63,51,65,62,59,62,63,68,64,67,60,59,59,61,58,65,64,70,63,68,62,61
M	70,70,50,68,71,65,70,73,70,69,64,68,65,72,73,75,72,75,64,69,60,68,66,69,72

PROBLEM 2-4

A research group in the military decided to test a new fabric for making uniforms. However, they realized that if they made two pairs of pants with differing material, only one pair could be worn at a time. Even if they restricted the testing to just one subject, variations could occur due to changes in that person's daily activities. They knew that the problem would get much worse as the testing expanded to more subjects, because variability between people would be severe. The experimenters overcame this problem by making special pants with the current fabric for one leg and the new fabric for the other leg. Conceivably, one leg might consistently get worn more (probably the right one), so they alternated which material got sewn on which leg.

They randomly assigned these special pants to nine subjects. (No problem finding "volunteers" in the military!) The results are shown in Table 2-8. The response is a subjective rating of wear on a scale of 1 to 10, the higher the better, assessed by a panel of inspectors and then averaged.

Table 2-8. Wear ratings for differing fabric for military pants

Block (Subject)	Old Fabric	New Fabric
1	8.1	9.0
2	5.1	5.8
3	6.8	7.2
4	8.5	8.8
5	6.7	7.6
6	4.4	4.3
7	6.4	6.9
8	6.7	7.3
9	5.8	6.1

No two subjects participated in exactly the same activities; however; both materials for any one person endured the same activities. By blocking the experiments by subject, the person-to-person variability due to differing activities (wear conditions) can be set aside during the analysis. Your mission is to determine whether the fabric wears differently, and if so, which one lasts longest. (Suggestion: set this up as a one-factor, blocked design. The ANOVA, if done properly, will remove the block variance before doing the F-test on the effect of the fabric. If the test is significant, make an effects plot.)

CHAPTER 3
TWO-LEVEL FACTORIAL DESIGN

"If you do not expect the unexpected, you will not find it."
—Heraclitus

If you've completed chapters 1 and 2 (or already mastered these basics), you're now equipped with very powerful tools to analyze experimental data. However, thus far we've restricted discussion to simple, comparative one-factor designs. We now introduce "factorial design"—a tool that allows you to experiment on many factors simultaneously. This chapter is arranged by increasing level of statistical detail. The latter portion becomes more mathematical, but the added effort required to study these details will pay off in increased understanding of the statistical framework and more confidence when using this powerful tool.

The simplest factorial design involves two factors each at two levels. The top part of Figure 3-1 shows the layout of this two-by-two design, which forms the square "X-space" on the left, versus the equivalent one-factor-at-a-time (OFAT) experiment at the upper right.

The points for the factorial designs are labeled in a "standard order," starting with all low levels and ending with all high levels. For example, runs 2 and 4 represent factor A at the high level. The average response from these runs can be contrasted with those from runs 1 and 3 (where factor A is at the low level) to determine the effect of A. Similarly, the top runs (3 and 4) can be contrasted with the bottom runs (1 and 2) for an estimate of the effect of B.

Later we will examine the mathematics of estimating effects, but the point to be made now is that a factorial design provides contrasts of averages, thus providing statistical power to the effect estimates. The OFAT experimenter must replicate runs to provide equivalent power. The end result for a two-

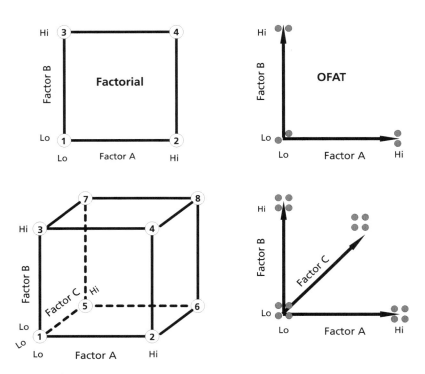

Figure 3-1. Two-level factorial versus one-factor-at-a-time (OFAT)

factor study is that to get the same precision for effect estimation, OFAT requires 6 runs, versus only 4 for the two-level design.

The advantage of factorial design becomes more pronounced as you add more factors. For example, with three factors, the factorial design requires only 8 runs (in the form of a cube), versus 16 for an OFAT experiment with equivalent power. In both designs (shown at the bottom of Figure 3-1), the effect estimates are based on averages of 4 runs each: right-to-left, top-to-bottom, and back-to-front for factors A, B, and C, respectively. The relative efficiency of the factorial design is now twice that of OFAT for equivalent power. The relative efficiency of factorials continues to increase with every added factor.

Factorial design offers two additional advantages over OFAT:

- Wider inductive basis, i.e., it covers a broader area or volume of X-space from which to draw inferences about your process.

- It reveals "interactions" of factors. This often proves to be the key to understanding a process, as you will see in the following case study.

Two-Level Factorial Design—As Simple as Making Microwave Popcorn

We will illustrate the basic principles of two-level factorial design via an example.

What could be simpler than making microwave popcorn? Unfortunately, it's nearly impossible to get every kernel of corn to pop. Often there's a considerable number of inedible "bullets" at the bottom of the bag. What causes this loss of popcorn yield? Think this over. The next time you stand in front of the microwave waiting for the popping to stop, jot down a list of all the possible factors affecting yield. You should easily identify five or even ten variables on your own, plus many more if you gather several colleagues or household members for a "brainstorm."

In our example below, only three factors were studied: brand of popcorn, time of cooking, and microwave power setting (see Table 3-1). The first factor, brand, is clearly "categorical"—either one type or the other. The second factor, time, is obviously "numerical," because it can be adjusted to any level. The third factor, power, could be set to any percent of the total available, so it's also numerical. If you try this experiment at home, be very careful to do some range finding on the high level for time (see related sidebar). Notice that we've introduced the symbols of minus (–) and plus (+) to designate low and high levels, respectively. This makes perfect sense for numerical factors, provided you do the obvious and make the lesser value the low level. However, the symbols for categorical factor levels are arbitrary.

Table 3-1. Test-factors for making microwave popcorn

Factor	Name	Units	Low Level (–)	High Level (+)
A	Brand	Cost	Cheap	Costly
B	Time	Minutes	4	6
C	Power	Percent	75	100

> ### BE AGGRESSIVE IN SETTING FACTOR LEVELS,
> ### BUT DON'T BURN THE POPCORN!
>
> One of the most difficult decisions for DOE, aside from which factors to chose, is at what levels to set them. A general rule is to set levels as far apart as possible so you will more likely see an effect, but not exceed the operating boundaries. For example, test pilots try to push their aircraft to the limits, called the "envelope." You don't want to break the envelope because the outcome may then be "crash and burn." In the actual experiment on popcorn (upon which the text example is loosely based), the experiment designer (one of the authors) set the upper level of time too high. In the randomized test plan, several other combinations were tested successfully before encountering the combination of high time and high power. Near the end of this run the popcorn erupted like a miniature volcano, emitting a lava-hot plasma of butter, steam and smoke. Alerted by the kitchen smoke alarm, the family gathered to observe the smoldering microwave oven. The author was heartened to hear the children telling his spouse not to worry because "in science, you learn from your mistakes." The spouse's reaction was not positive, but a new microwave restored harmony to the household. The author now advises that as a safety precaution you conduct a highly controlled pre-trial on the extreme combination(s) of factors.

Two responses were considered for the experiment on microwave popcorn: taste and "bullets." Taste was determined by a panel of testers who rated the popcorn on a scale of 1 (worst) to 10 (best). The ratings were averaged and multiplied by 10. This is a linear "transformation" that eliminates a decimal point to make data entry and analysis easier. It does not affect the relative results. The second response, "bullets," was measured by weighing the unpopped kernels—the lower the better.

The results from doing all combinations of the chosen factors each at two levels are shown in Table 3-2. Taste ranged from a 32 to 81 rating, and "bullets" from 0.7 to 3.5 ounces. The latter result came from a bag with virtually no popped corn—barely enough to even get a taste. Obviously, this particular setup is one to avoid. The run order was randomized to offset any lurking variables, such as machine warm-up and degradation of taste buds.

TWO-LEVEL FACTORIAL DESIGN

> ### ALWAYS RANDOMIZE YOUR RUN ORDER
> You must randomize the order of your experimental runs to satisfy the statistical requirement of independence of observations. Randomization acts as insurance against the effects of lurking time-related variables, such as the warm-up effect on a microwave oven. For example, let's say you forget to randomize and first run all low levels of a factor and then all high levels of a given factor that actually creates no effect on response. Meanwhile, an uncontrolled variable causes the response to gradually increase. In this case, you will mistakenly attribute the happenstance effect to the non-randomized factor. By randomizing the order of experimentation, you greatly reduce the chances of such a mistake. Select your run numbers from a table of random numbers or mark them on slips of paper and simply pull them blindly from a container. Statistical software can also be used to generate random run orders.

The first column in Table 3-2 lists the standard order, which can be cross-referenced to the labels on the three-factor cube in Figure 3-2. We also placed the mathematical symbols of minus and plus, called "coded factor levels," next to the "actual" levels at their lows and highs, respectively. To take advantage of established methods for analysis, it will be very helpful to re-sort the test matrix on the basis of standard order, and list only the coded factor levels. We also want to dispense with the names of the factors and responses, which just get in the way of the calculations, and show only their mathematical symbols. You can see the results in Table 3-3.

Table 3-2. Results from microwave popcorn experiment

Standard Order	Run Order	A: Brand	B: Time (minutes)	C: Power (percent)	Y_1: Taste (rating)	Y_2: "bullets" (ounces)
2	1	Costly (+)	4 (−)	75 (−)	75	3.5
3	2	Cheap (−)	6 (+)	75 (−)	71	1.6
5	3	Cheap (−)	4 (−)	100 (+)	81	0.7
4	4	Costly (+)	6 (+)	75 (−)	80	1.2
6	5	Costly (+)	4 (−)	100 (+)	77	0.7
8	6	Costly (+)	6 (+)	100 (+)	32	0.3
7	7	Cheap (−)	6 (+)	100 (+)	42	0.5
1	8	Cheap (−)	4 (−)	75 (−)	74	3.1

DOE SIMPLIFIED

Table 3-3. Test matrix in standard order with coded levels

Standard	Run	A	B	C	Y_1	Y_2
1	8	−	−	−	74	3.1
2	1	+	−	−	75	3.5
3	2	−	+	−	71	1.6
4	4	+	+	−	80	1.2
5	3	−	−	+	81	0.7
6	5	+	−	+	77	0.7
7	7	−	+	+	42	0.5
8	6	+	+	+	32	0.3
Effect Y_1		−1.0	−20.5	−17.0	66.5	
Effect Y_2		−0.05	−1.1	−1.8		1.45

The column labeled "Standard," plus the columns for A, B, and C form a template that can be used for any three factors that you want to test at two levels. The standard layout starts with all minus (low) levels of the factors and ends with all plus (high) levels. The first factor changes sign every other row; the second factor every second row; the third every fourth row; and so on, based on powers of 2. You can extrapolate the pattern to any number of factors, or look them up in statistical handbooks.

ORTHOGONAL ARRAYS [WHEN YOU HAVE LIMITED RESOURCES, IT PAYS TO PLAN AHEAD]

The standard two-level factorial layout shown in Table 3-3 is one example of a carefully balanced "orthogonal array." Technically, this means that there's no correlation among the factors. You can see this most clearly by looking at column C. When C is at the minus level, factors A and B contain an equal number of pluses and minuses; thus, their effect cancels. The same result occurs when C is at the plus level. Therefore, the effect of C is not influenced by factors A or B. The same can be said for the effects of A and B and all the interactions as well. We're sticking with what are commonly called the standard arrays for two-level full and fractional factorials. However, you may come across other varieties of orthogonal arrays, such as Taguchi and Plackett-Burman. Any orthogonal test array is much preferred to unplanned experimentation (an oxymoron). Happenstance data is likely to be highly correlated (nonorthogonal), which makes it much more difficult to sort out the factors that really affect your response.

TWO-LEVEL FACTORIAL DESIGN

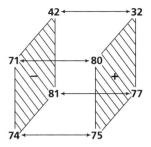

Figure 3-2. Cube plot of taste ratings with focus on brand (Factor A)

Let's begin the analysis by investigating the "main effects" on the first response (Y_1)—taste. It helps to view the results in the cubical factor space. We will focus on factor A (brand) first.

The right side of the cube contains all the runs where A is at the plus level (high), versus the left side, where the factor is held at the minus level (low). Now simply average the highs and the lows, and determine the difference or contrast: this is the effect of factor A. Mathematically, the calculation of an effect is expressed as follows:

$$\text{Effect} = \frac{\sum Y_+}{n_+} - \frac{\sum Y_-}{n_-}$$

where the n's refer to the number of data points you've collected at each level. The Y's refer to the associated responses. You can pick these off the plot or from the matrix itself. For A, the effect is:

$$\text{Effect} = \frac{75 + 80 + 77 + 32}{4} - \frac{74 + 71 + 81 + 42}{4} = 66 - 67 = -1$$

In comparison to the overall spread of results, it looks like A (brand) has very little effect on taste. Continue the analysis by contrasting the averages from top-to-bottom and back-to-front to get the effects of B and C, respectively. Go ahead and do the calculations if you like. The results are –20.5 for B and –17 for C. The impact, or "effect," of factors B (power) and C (time) are much larger than that of A (the brand of popcorn).

However, before you jump to conclusions, we must consider the effects caused by interactions of factors. The full-factorial design allows estimates of all three two-factor interactions, AB, AC, and BC, plus the three-factor interaction, ABC. Including the main effects (caused by A, B, and C), that brings the total to seven effects—the most you can estimate from the eight-run factorial design, because one degree of freedom is used to estimate the overall mean.

Table 3-4 lists all seven effects. The main effects calculated earlier are listed under the A, B, and C columns.

The pattern of pluses and minuses for interaction effects are calculated by multiplying the parent terms. For example, the AB column is the product of the A and B columns, so for the first standard row, the combination of –A times –B produces a +AB. Remember that like signs, when multiplied, produce a plus, whereas the multiplication of minus and plus or vice versa makes a minus. The entire array exhibits a very desirable property of balance called "orthogonality" (see related sidebar).

Now it's just a matter of computing the effects using the general formula shown previously. The results are shown on the bottom line of Table 3-4. Notice that the interaction effect of BC is even larger on an absolute scale than its parents, B and C. In other words, the combination of time (B) and power (C) produces a big (negative) impact on taste. With that as a clue, look

Table 3-4. Complete matrix, including interactions, with effects calculated

Standard	Main Effects			Interaction Effects				Response
	A	B	C	AB	AC	BC	ABC	Y_1
1	−	−	−	+	+	+	−	74
2	+	−	−	−	−	+	+	75
3	−	+	−	−	+	−	+	71
4	+	+	−	+	−	−	−	80
5	−	−	+	+	−	−	+	81
6	+	−	+	−	+	−	−	77
7	−	+	+	−	−	+	−	42
8	+	+	+	+	+	+	+	32
Effect	-1.0	-20.5	-17.0	0.5	-6.0	-21.5	-3.5	66.5

TWO-LEVEL FACTORIAL DESIGN

more closely at the response data (Y_1). Notice the big drop-off in taste when both B and C are at their high levels. We'll investigate this further, after sorting out everything else.

On an absolute value scale, the other interaction effects range from near 0 (for AB) to as high as 6 (for AC). Could these just be chance occurrences due to normal variations in the popcorn, the tasting, the environment, and the like? To answer this question, let's go back to a tool discussed at the end of Chapter 1: the normal plot. Then we can see whether some or all of the effects vary normally. In this case, we hope to see one or more effects at a significant distance from the remainder. Otherwise we've wasted a lot of experimental effort chasing noise from the system.

Before plotting the effects, it helps to convert them to absolute values, a more sensitive scale for detection of significant outcomes. The absolute value scale is accommodated via a variety of normal paper called the "half-normal," which is literally based on the positive half of the full normal curve. (Imagine cutting out the bell-shaped curve and folding it in half at the mean.) As before, the vertical (Y) axis of the half-normal plot displays the cumulative probability of getting a result at or below any given level. However the probability scale for the half-normal is adjusted to account for using the absolute value of the effects.

Remember that before plotting this data on the probability paper you must:

1. Sort the data points (in this case 7 effects) in ascending order.
2. Divide the 0 to 100% cumulative probability scale into (7) equal segments.
3. Plot the data at the midpoint of each probability segment.

In this case, each probability segment will be approximately 14.28% (100/7). The lowest weight will be plotted at 7.14%, which is the midpoint of the first segment. Table 3-5 shows this combination and all the remaining ones.

Now all we need to do is plot the absolute values of the effect on the x-axis versus the cumulative probabilities on the specially scaled y-axis on half-normal paper (see Figure 3-3). In ancient times (before personal computers), you would get this from a commercial vendor of graph paper, but now most people generate half-normal plots from statistical software.

DOE SIMPLIFIED

Table 3-5. Values to plot on half-normal probability paper

Point	Effect	Absolute Value of Effect	Cumulative Probability
1	AB	|0.5|	7.14%
2	A	|–1.0|	21.43%
3	ABC	|–3.5|	35.71%
4	AC	|–6.0|	50.00%
5	C	|–17.0|	64.29%
6	B	|–20.5|	78.57%
7	BC	|–21.5|	92.86%

Figure 3-4 shows the completed half-normal plot for the effects on taste of popcorn. This particular graph has some added features you will not usually see:

- A half-normal curve for reference.
- A "dot-plot" on the x-axis representing the actual effects projected down to the x-axis number line.

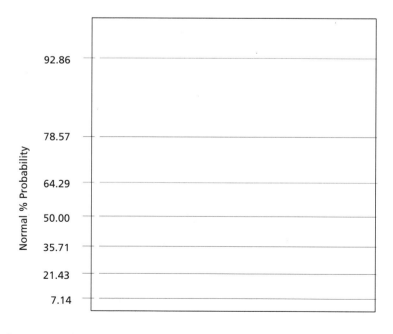

Figure 3-3. Blank half-normal paper (set up for plotting 7 effects)

TWO-LEVEL FACTORIAL DESIGN

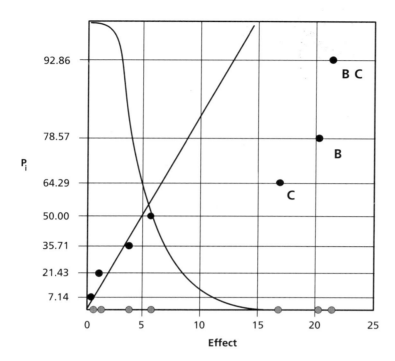

Figure 3-4. Half-normal plot of effects for taste (curve and dot-plot added for reference)

Notice that the biggest three effects fall well out on the tail of the normal curve (to the right). These three effects (C, B, and BC) are most likely significant in a statistical sense. We wanted to draw attention to these big effects, so we labeled them. Observe the large gap before you get to the next lowest effect. From this point on the effects (AC, ABC, A, and AB—from biggest to smallest, respectively) fall in line, which represents the normal scatter. We deliberately left these unlabeled to downplay their importance. These four trivial effects (nearest 0) will be used later as an estimate of error for analysis of variance (ANOVA).

The pattern you see in Figure 3-4 is very typical: the majority of points fall in a line emanating from the origin, followed by a gap, and then one or more points fall off to the right of the line. The half-normal plot of effects makes it very easy to see at a glance what's significant, if anything.

> ### THE VITAL FEW VERSUS THE TRIVIAL MANY
>
> A rule of thumb, called "sparsity of effects," says that in most systems, only 20% of the main effects ("ME") and two-factor interactions ("2 fi") will be significant. The other ME and 2 fi's, as well as any three-factor interactions ("3 fi") or greater will vary only to the extent of normal error. (Remember that the effects are based on averages, so their variance will be reduced by a factor of n.) This rule-of-thumb is very similar to that developed a century ago by economist Vilfredo Pareto, who found that 80% of the wealth was held by 20% of the people. Dr. Joseph Juran, a pre-eminent figure in the 20th century quality movement, applied this 80/20 rule to management: 80% of the trouble comes from 20% of the problems. He advised that you focus your efforts on these "vital few" problems and ignore the "trivial many."

Let's apply this same procedure to the second response for microwave popcorn—the weight of the "bullets." In the last row of Table 3-6, the seven effects are calculated using the formula shown earlier:

$$\text{Effect} = \frac{\sum Y_+}{n_+} - \frac{\sum Y_-}{n_-}$$

Table 3-7 shows the effects ranked from low to high in absolute value, with the corresponding probabilities.

Table 3-6. Effects calculated for second response (bullets)

Standard	A	B	C	AB	AC	BC	ABC	Y_2
1	−	−	−	+	+	+	−	3.1
2	+	−	−	−	−	+	+	3.5
3	−	+	−	−	+	−	+	1.6
4	+	+	−	+	−	−	−	1.2
5	−	−	+	+	−	−	+	0.7
6	+	−	+	−	+	−	−	0.7
7	−	+	+	−	−	+	−	0.5
8	+	+	+	+	+	+	+	0.3
Effect	−0.05	−1.1	−1.8	−0.25	−0.05	0.80	0.15	1.45

TWO-LEVEL FACTORIAL DESIGN

Table 3-7. Values to plot on half-normal plot for bullets

Point	Effect	Absolute Value of Effect	Cumulative Probability
1	A	|−0.05|	7.14%
2	AC	|−0.05|	21.43%
3	ABC	|0.15|	35.71%
4	AB	|−0.25|	50.00%
5	BC	|0.80|	64.29%
6	B	|−1.10|	78.57%
7	BC	|−1.80|	92.86%

Notice that the probability values are exactly the same as for the previous table on taste. In fact, these values apply to any three-factor, two-level design, so long as you successfully perform all eight runs and gather the response data.

Figure 3-5 shows the resulting plot (computer generated) for "bullets," with all effects labeled so you can see how it's constructed. For example, the smallest effects, A and AC, which each have an absolute value of 0.05, are plotted at 7.1 and 21.4% probability. (When effects are equal, the order is arbitrary.) Next comes effect ABC at 35.7%, and so on.

Notice how four of the effects (AB, ABC, AC, and A) fall in a line near zero. Assume that these effects vary only due to normal causes and consider them to be insignificant. You will almost always find three-factor interactions, such as ABC, in this normal population of trivial effects. Interactions of four or more factors are even more likely fall in this near-zero group of effects.

The effects of B, C, and BC are very big relative to the other effects. They obviously do not fall on the line. In a statistical sense, each of these three standout effects should be considered significant populations in their own right. In other words, we need to focus on factors B and C and how they interact (as BC) to affect the response of "bullets."

To protect against spurious outcomes, it is absolutely vital that you verify the conclusions drawn from the half-normal plots by doing an analysis of variance (ANOVA) and the associated diagnostics of "residual error." As you will see later in this chapter, the statistics in this case pass the tests with flying colors. Please take our word on it for now: we will eventually show you how to

DOE SIMPLIFIED

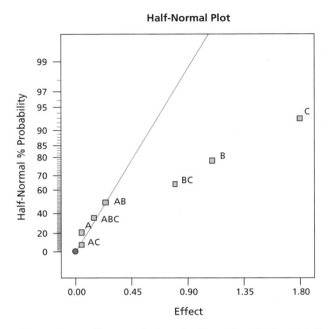

Figure 3-5. Half-normal plot of effects for "bullets" (all effects labeled)

generate and interpret all the statistical details, but we thought it might be more interesting now to jump ahead to the effect plots.

How to Plot and Interpret Interactions

Interactions occur when the effect of one factor depends on the level of the other. They cannot be detected by one traditional one factor at a time (OFAT) experimentation, so don't be surprised if you uncover previously undetected interactions when you run a two-level design. The result very often will be a breakthrough improvement in your system.

The microwave popcorn study nicely illustrates how to display and interpret an interaction. In this case, both of the measured responses are greatly impacted by the interaction of time and power, so it is helpful to focus on these two factors (B and C, respectively). Table 3-8 shows the results for the two responses: taste and "bullets." These are actually averages of data from Table 3-3, which we've cross-referenced by standard order. For example, the

TWO-LEVEL FACTORIAL DESIGN

Table 3-8. Data for interaction plot of microwave time versus power

Standard	Time (B)	Power (C)	Taste (Y$_1$ Avg.)	"Bullets" (Y$_2$ Avg.)
1,2	–	–	74.5	3.3
3,4	+	–	75.5	1.4
5,6	–	+	79.0	0.7
7,8	+	+	37.0	0.4

first two experiments in Table 3-3 have both time and power at their low (minus) levels. The associated taste ratings are 74 and 75, which produce an average outcome of 74.5, as shown in Table 3-4.

Notice that the effect of time depends on the level of power. For example, when power is low (minus), the change in taste is small—from 74.5 to 75.5. But when power is high (plus), the taste goes very bad—from 79 to 37. This is much clearer when graphed (see Figure 3-6).

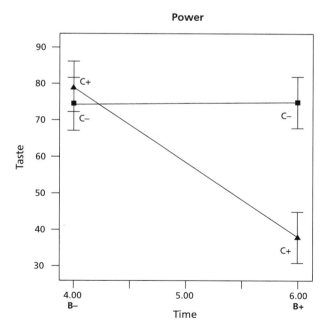

Figure 3-6. Interaction of time (B) versus power (C) on popcorn taste

Two lines appear on the plot, bracketed by least significant difference (LSD) bars at either end. The lines are far from parallel, indicating quite different effects of changing the cooking time. When power is low (C–), the line is flat, which indicates that the system is unaffected by time (B). But when power goes high (C+), the line angles steeply downward, indicating a strong negative effect due to the increased time. The combination of high time and high power is bad for taste. Table 3-8 shows the average result to be only 37 on the 100-point rating scale. The reason is simple: the popcorn burns. The solution to this problem is also simple: turn off the microwave sooner. Notice that when the time is set at its low level (B–), the taste remains high regardless of the power setting (C). The LSD bars overlap at this end of the interaction graph, which implies that there is no significant difference in taste.

TASTE IS IN THE MOUTH OF THE BEHOLDER

Before being re-scaled, the popcorn taste was rated on a scale of 1 (best) to 10 (worst) by a panel of Minnesotans. The benchmarks they used reflect a conservative, Scandinavian heritage:

10 – Just like ludefisk*	5 – Could be worse
9 – Not bad for you	4 – My spouse made something like this once
8 – Tastes like Mom's	3 – I like it , but…
7 – Not so bad	2 – It's different
6 – Could be better	1 – Complete silence

*(Fish preserved by being soaked in lye. Also good for removing wallpaper.)

If you're not from Minnesota, we advise that you use an alternative scale used by many sensory evaluators, which goes from 1 to 9, with 9 being the best. All nine numbers are laid out in line. The evaluator circles the number that reflects their rating of a particular attribute. To avoid confusion about orientation of the scale, we advise that you place sad (☹), neutral (😐), and happy (☺) faces at the 1, 5, and 9 positions on the number line, respectively. This is called a "hedonic" scale. Rating scales like this can provide valuable information on subjective responses, particularly when you apply the averaging power of a well-planned experiment.

TWO-LEVEL FACTORIAL DESIGN

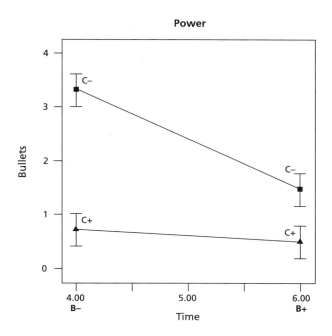

Figure 3-7. Interaction of time (B) versus power (C) on popcorn "bullets"

Figure 3-7 shows how time and power interact to affect the other response, the "bullets." The pattern differs from that for taste, but it again exhibits the nonparallel lines that are characteristic of a powerful two-factor interaction.

The effect of time on the weight of "bullets" depends on the level of power, represented by the two lines on the graph. On the lower line, notice the overlap in the least significant difference (LSD) bars at left versus right. This indicates that at high power (C+), there's not much, if any, effect. However, the story differs for the top line on the graph where power is set at its low level (C–). Here the LSD bars do not overlap, indicating that the effect of time is significant. Getting back to that bottom line, it's now obvious that, using the "bullets" as a gauge, it's best to make microwave popcorn at the highest power setting. However, recall that high time and high power resulted in a near-disastrous loss of taste. Therefore, for "multiresponse optimization" of the microwave popcorn, the best settings are high power at low time. The brand, factor A, does not appear to significantly affect either response, so choose the one that's cheapest.

Protect Yourself with Analysis of Variance (ANOVA)

Now that we've jumped to conclusions on how to make microwave popcorn, it's time to do our statistical homework by performing the analysis of variance (ANOVA). Fortunately, when factorials are restricted to two levels, the procedure becomes relatively simple. We've already done the hard work by computing all the effects. To do the ANOVA, we must compute the sums of squares (SS), which are related to the effects as follows:

$$SS = \frac{N}{4} \text{Effect}^2$$

N is the number of runs. This formula works only for balanced two-level factorials.

The three largest effects, B, C, and BC, are the vital few that stood out on the half-normal plot. Their sum of squares is shown in the italicized rows in Table 3-9. For example, the calculation for sum of squares for effect B is:

$$SS_B = \frac{8}{4}(-20.5)^2 = 840.5$$

You can check the calculations for the sum of squares associated with effects C and BC. The outstanding effects will be incorporated in the "model" for predicting the taste response. (We'll provide more details on the model later.)

When added together, the resulting sum of squares provides the beginnings of the actual ANOVA. Here's the calculation for the taste response:

$$SS_{Model} = SS_B + SS_C + SS_{BC} = 840.5 + 578 + 924.5 = 2343$$

The smaller effects, which fell on the near-zero line, will be pooled together and used as an estimate of error called "residual." Here's the calculation for the taste response:

$$\begin{aligned}SS_{Residual} &= SS_A + SS_{AB} + SS_{AC} + SS_{ABC} \\ &= \frac{8}{4}(-1)^2 + \frac{8}{4}(-0.5)^2 + \frac{8}{4}(-6)^2 + \frac{8}{4}(-3.5)^2 \\ &= 0.5 + 2 + 72 + 24.5 = 99\end{aligned}$$

TWO-LEVEL FACTORIAL DESIGN

Table 3-9. ANOVA for taste

Source	Sum of Squares (SS)	Df	Mean Square (MS)	F Value	Prob >F
Model	2343.0	3	781.0	31.5	<0.01
B	840.5	1	840.5	34.0	<0.01
C	578.0	1	578.0	23.3	<0.01
BC	924.5	1	924.5	37.3	<0.01
Residual	99.0	4	24.8		
Cor Total	2442.0	7			

The sum of squares for model and residual are shown in the first column of data in the ANOVA shown in Table 3-9. The next column lists the degrees of freedom (df) associated with the sum of squares (derived from the effects). Each effect is based on two averages, high versus low, so it contributes 1 degree of freedom (df) for the sum of squares. Thus, you will see 3 df for the three effects in the model pool and 4 df for the four effects in the residual pool. This is another simplification made possible by restricting the factorial to two levels. The next column in the ANOVA is the mean square: the sum of squares divided by the degrees of freedom (SS/df). The ratio of mean squares ($MS_{Model}/MS_{Residual}$) forms the F value of 31.5 (=781.0/24.8).

The F value for the model must be compared to the reference distribution for F with the same degrees of freedom. In this case, you've got 3 df for the numerator (top) and 4 df for the denominator (bottom). The critical F-values can be attained from table(s) in the Appendix by going to the appropriate column (in this case the third) and row (the fourth). Check these against the values shown in Figure 3-8.

If the actual F value exceeds the critical value at an acceptable risk value, you should reject the null hypothesis. In this case, the actual F of 31.5 is bracketed by the critical values for 0.1% and 1% risk. We can say that the probability of getting an F as high as that observed, due to chance alone, is less than 1%. In other words, we are more than 99% confident that taste is significantly affected by one or more of the effects chosen for the model. That's good!

But we're not done yet, because it's possible to accidentally carry an insignificant effect along for the ride on the model F. For that reason, always check

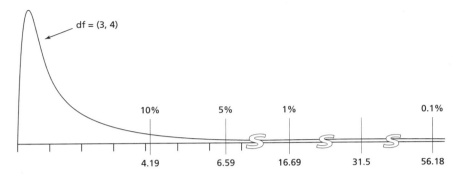

Figure 3-8. The F-distribution with various critical values noted

each individual effect for significance. The F-tests for each effect are based on 1 df for the respective numerators and the df of the residual for the denominator (in this case, 4). The critical F at 1% for these df (1 and 4) is 21.2. Check the appropriate table in the Appendix to verify this. The actual F-values for all three individual effects exceed the critical F, so we can say they're all significant, which supports the assumptions made after viewing the half-normal plot.

We haven't talked about the last line of the ANOVA, labeled "Cor Total." This is the total sum of squares corrected for the mean. It represents the total system variation using the average response as a baseline. The degrees of freedom are also summed, so you can be sure nothing is overlooked. In this case we started with eight data points, but 1 df is lost to calculate the overall mean, leaving seven df for the ANOVA.

The ANOVA for the second response, "bullets," can be constructed in a similar fashion. The one shown in Table 3-10 is from a computer program that calculates the probability (p) value to several decimals (reported as "Prob > F"). The p-values are traditionally reported on a scale from 0 to 1. In this book, p-values less than 0.05 are considered significant, providing at least 95% confidence for all results. None of the p-values exceed 0.05 (or even 0.01), so we can say that the overall model for "bullets" is significant, as are the individual effects.

TWO-LEVEL FACTORIAL DESIGN

Table 3-10. ANOVA for "bullets"

Source	Sum of Squares	Df	Mean Square	F Value	Prob >F
Model	10.18	3	3.39	75.41	0.0006
B	2.42	1	2.42	53.78	0.0018
C	6.48	1	6.48	144.00	0.0003
BC	1.28	1	1.28	28.44	0.0006
Residual	0.18	4	0.045		
Cor Total	10.36	7			

Let's recap the steps taken so far for analyzing two-level factorial designs:

1. Calculate effects—average of highs (pluses) versus average of lows (minuses).
2. Sort absolute value of effects in ascending order.
3. Calculate probability values, P_i, using formula.
4. Plot effects on half-normal probability paper.
5. Fit line through near-zero points ("residual").
6. Label significant effects off the line ("model").
7. Calculate the sums of squares (SS) using formula.
8. Compute SS_{Model}: Add SS for points far from line.
9. Compute $SS_{Residuals}$: Add SS for points on line.
10. Construct ANOVA table.
11. Using tables, estimate the p-values for calculated F-values. If <0.05, proceed.
12. Plot main effect(s) and interaction(s). Interpret results.

We've now completed most of the statistical homework needed to support the conclusions made earlier. However, there's one more very important step needed for absolute protection: check the assumptions underlying the ANOVA.

Modeling Your Responses with Predictive Equations

This is a good place to provide details on the model tested in the ANOVA. The model is a mathematical equation used to predict a given response. To keep it simple, let's begin the discussion by looking at only one factor. The linear model is:

$$\hat{Y} = \beta_0 + \beta_1 X_1$$

where Y with 'hat' (^) is the predicted response, β_0 (beta nought) is the intercept, and β_1 (beta one) is the model coefficient for the input factor (X_1). For statistical purposes, it helps to keep factors in coded form: −1 for low and +1 for high. As shown in Figure 3-9, changing the factor from low to high causes the measured effect on response.

The model coefficient β_1 represents the slope of the line, which is the "rise" in response (the effect) divided by the corresponding "run" in factor level (2 coded units). Therefore, the β_1 coefficient in the coded model is one-half the value of the effect (effect/2).

As more factors are added, the number of terms in the model increases. The factorial model for two factors, each at two levels, is:

$$\hat{Y} = \beta_0 + \beta_1 X_1 + \beta_2 X_2 + \beta_{12} X_1 X_2$$

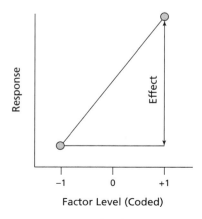

Figure 3-9. Graph of response versus coded factor level

Here's the model for the popcorn taste with the factors of time (B) and power (C) in coded form:

$$\text{Taste} = 66.5 - 10.25\,B - 8.50\,C - 10.75\,BC$$

The value for the intercept (β_0) of 66.5 represents the average of all actual responses. The coefficients can be directly compared to assess the relative impact of factors. In this case, for example, we can see that factor B (coefficient -10.25) causes a bigger effect than factor C (coefficient -8.50).

The one drawback to the coded model is that you must convert actual factor levels to coded levels before plugging in the input values. Using standard statistical regression, we produced an alternative predictive model that expresses the factors of time and power as their original units of measure:

$$\text{Taste} = -199 + 65\,\text{Time} + 3.62\,\text{Power} - 0.86\,\text{Time}*\text{Power}$$

Use this uncoded model to generate predicted values, but don't try to interpret the coefficients. The intercept loses meaning when you go to the uncoded model because it's dependent on units of measure. For example, a −199 result for taste makes no sense. Similarly, in the uncoded model you can no longer compare the coefficient of one term with another, such as time versus power.

We advise that you work only with the coded model. This is shown below for the second response:

$$\text{"bullets"} = 1.45 - 0.55\,B - 0.90\,C + 0.40\,BC$$

A good way to check your models is to enter factor levels from your design and generate the predicted response. When you compare the predicted value with the actual (observed) value, you will always see a discrepancy. This is called the "residual."

Diagnosing Residuals to Validate Statistical Assumptions

For statistical purposes it is assumed that residuals are normally distributed and independent with constant variance. Two plots are recommended for checking the statistical assumptions:

- Normal plot of residuals.
- Residuals versus predicted level.

Let's look at these plots for the taste response from the popcorn experiment. Table 3-11 provides the raw data.

The column of predicted ("Pred") values for taste is determined by plugging the coded factor levels into the coded model. For example, the predicted taste for standard order 1 is:

$$\text{Taste} = 66.5 - 10.25(-1) - 8.50(-1) - 10.75(+1) = 74.5$$

The residuals ("Resid"), calculated from the difference of actual versus predicted response, can be plotted on normal probability paper. The procedure for creating a *full*-normal plot is the same as that shown earlier for the *half*-normal plot, but you don't need to take the absolute value of the data. Just be sure you've got the correct variety of graph paper! In this case, we've got eight points (m = 8), so the P_i from the formula given earlier are 6.25, 18.75, 31.25, 43.75, 56.25, 68.75, 81.25, and 93.75 percent. The resulting plot is shown on Figure 3-10.

Table 3-11. Residuals for taste data

Standard	B	C	BC	Taste Actual	Taste Pred	Resid
1	−1	−1	+1	74	74.5	-0.5
2	−1	−1	+1	75	74.5	0.5
3	+1	−1	−1	71	75.5	-4.5
4	+1	−1	−1	80	75.5	4.5
5	−1	+1	−1	81	79.0	2.0
6	−1	+1	−1	77	79.0	-2.0
7	+1	+1	+1	42	37.0	5.0
8	+1	+1	+1	32	37.0	-5.0

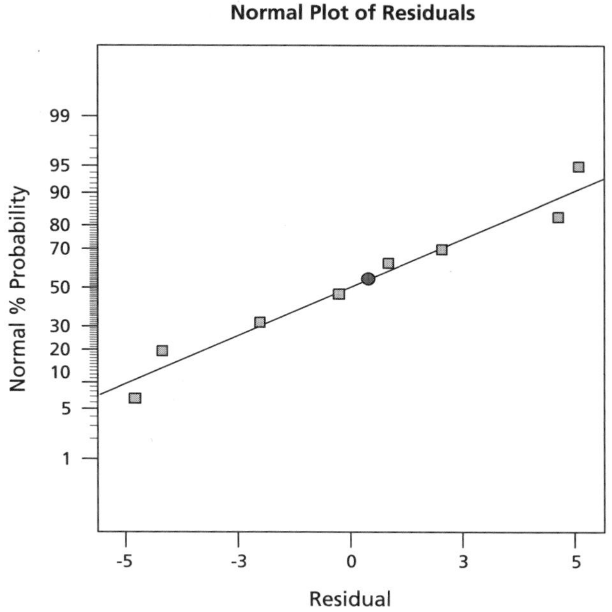

TWO-LEVEL FACTORIAL DESIGN

Normal Plot of Residuals

Figure 3-10. Normal plot of residuals for popcorn taste

If the residuals are normally distributed, they will all fall in a line on this special paper. In this case, the deviations from linear are very minor, so it supports the assumption of normality. Watch for clearly nonlinear patterns, such as an "S" shape. Then consider doing a response transformation—a topic that will be discussed in the next section.

THE "PENCIL TEST"

A simple way to check for linearity is to place a pencil on the graph. If the pencil covers all the points, consider it in line. A big marker-pen would solve all your problems!

Figure 3-11 shows the normal plot of residuals for the second response ("bullets"). Give it the "pencil test"! You will find that residuals for the bullets exhibit no major deviations from the normal line.

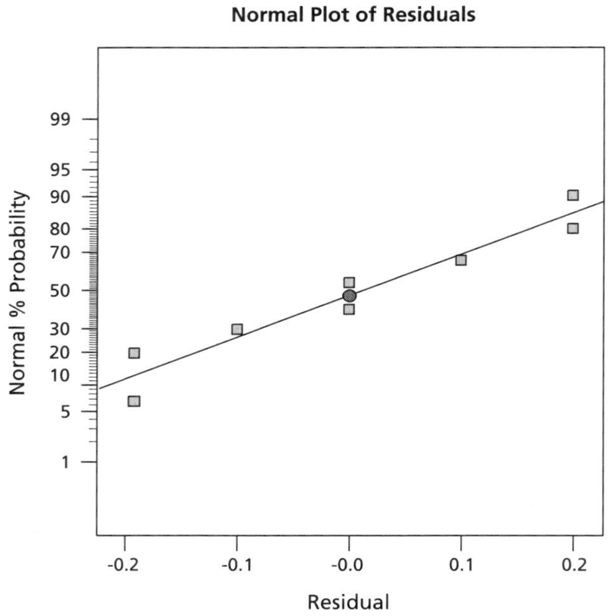

Normal Plot of Residuals

Figure 3-11. Normal plot of residuals for popcorn "bullets"

The other recommended plot for diagnostics is the residuals versus predicted response. Using the data from Table 3-11, we constructed the plot shown in Figure 3-12.

Ideally, the vertical spread of data will be approximately the same from left to right. Watch for a megaphone (<) pattern, where the residuals increase with the predicted level. In a design as small as that used for the popcorn experiment, only eight runs, it's hard to detect patterns. However, it's safe to say that there's no definite increase in residuals with predicted level, which supports the underlying statistical assumption of constant variance. In the next chapter we will show you what to do if the residuals are not normal and exhibit nonconstant variance.

TWO-LEVEL FACTORIAL DESIGN

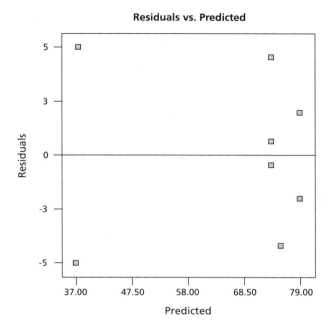

Figure 3-12. Residuals versus predicted taste

A LITERAL RULE-OF-THUMB

ANOVA and other statistical analyses are relatively robust to deviations from normality and constancy of variance. Therefore, you should not overreact to slight nonlinearity on the normal plot of residuals, or vague patterns on the residuals versus predicted plot. As a rule of thumb, if you think you see a pattern, but it disappears when you cover one point with your thumb, then don't worry. However, if you construct these plots on your computer, remember to wipe the thumb-print off your display.

P.S. Despite rumors to the contrary, the term "rule of thumb" probably came from use of the thumb as a crude measure of length. However, it may refer to the traditional practice of brewmasters to check temperature of beer by dipping their thumb in the batch. This latter explanation is most consistent with the origins of the t-test and other statistical innovations.

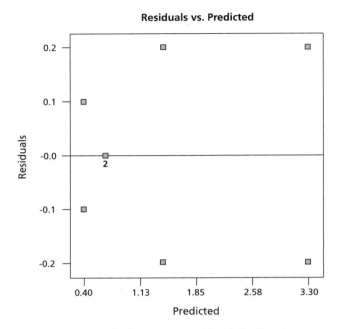

Figure 3-13. Residuals versus predicted "bullets"

Figure 3-13 shows the residual versus predicted plot for the second response ("bullets"). You don't need to apply the "rule of thumb" because there's no obvious increase in residuals as the predicted value increases.

Practice Problems

PROBLEM 3-1

Montgomery describes a two-level design on a high-pressure chemical reactor (see the referenced textbook, *Design and Analysis of Experiments*, example 7-2, page 319). A full-factorial experiment is carried out in the pilot plant to study four factors thought to influence the filtration rate of the product. Table 3-12 shows actual high and low levels for each of the factors.

At each combination of these machine settings, the experimenters recorded the filtration rate. The goal is to maximize the filtration rate and also try to

TWO-LEVEL FACTORIAL DESIGN

Table 3-12. Factors and levels for two-level factorial design on a reactor

Factor	Name	Units	Low Level (–)	High Level (+)
A	Temperature	Deg C	24	35
B	Pressure	PSIG	10	15
C	Concentration	Percent	2	4
D	Stir Rate	RPM	15	30

find conditions that would allow a reduction in the concentration of formaldehyde, factor C.

The response data are tabulated in standard order, with factor levels coded, in Table 3-13.

Do an analysis of the data to see if any effects are significant. Recommend operating conditions that maximize rate with a minimum of formaldehyde.

Table 3-13. Design layout and response data for reactor study

Standard	A	B	C	D	Filtration Rate (gallons per hour)
1	–	–	–	–	45.0
2	+	–	–	–	71.0
3	–	+	–	–	48.0
4	+	+	–	–	65.0
5	–	–	+	–	68.0
6	+	–	+	–	60.0
7	–	+	+	–	80.0
8	+	+	+	–	65.0
9	–	–	–	+	43.0
10	+	–	–	+	100.0
11	–	+	–	+	45.0
12	+	+	–	+	104.0
13	–	–	+	+	75.0
14	+	–	+	+	86.0
15	–	+	+	+	70.0
16	+	+	+	+	96.0

(Suggestion: use the software provided with the book. First do the two-level factorial tutorial that comes with the program. It's keyed to the data in Table 3-12. See the accompanying insert for software installation instructions and details on the associated tutorials.)

PROBLEM 3-2

Modern cars are built with such precision that they become hermetically sealed when locked. As a result, the interior becomes unbearably hot in cars parked outdoors on warm, sunny days. A variety of window covers can be purchased to alleviate the heat. The materials vary, but generally the covers present either a white or shiny, metallic surface that reflects solar radiation. In some cases, they can be flipped to one side or the other. The white variety of cover usually displays some sort of printed pattern, such as a smiling sun or the logo of a local sports team. These patterns look good, but they may detract from the heat-shielding effect. A two-level factorial design was conducted to quantify the effects of several potential variables: cover (shiny versus white), direction of the parked car (east versus west), and location (close by the office in an open lot versus far away from the office under a shade tree).

> **FOR THOSE CONSUMERS WHO MAY NOT BE FIRING ON ALL CYLINDERS**
> Operating instructions seen on accordion-style front-window shade for automobiles: "Remove before driving."

The resulting eight-run, two-level DOE was performed during a period of stable weather in Minneapolis during early September. However, just in case variations did occur, the experimenter recorded temperature, cloudiness, wind speed, and other ambient conditions. Outside temperatures ranged from 66 to 76 degrees Fahrenheit under generally clear skies. Randomization of the run order provided insurance against the minor variations in weather. The response shown in Table 3-14 is the difference in temperature from inside to outside, as measured by a digital thermometer.

Even in this relatively mild late-summer season, the inside temperatures of the automobile often exceeded 100 degrees Fahrenheit. It's not hard to imagine how hot it could get under extreme midsummer weather. Analyze this

Table 3-14. Results from car-shade experiment

Standard	A: Cover	B: Orientation	C: Location	Temp Increase (Deg F)
1	White	East	Close/Open	42.1
2	Shiny	East	Close/Open	20.8
3	White	West	Close/Open	54.3
4	Shiny	West	Close/Open	23.2
5	White	East	Far/Shaded	17.4
6	Shiny	East	Far/Shaded	10.4
7	White	West	Far/Shaded	11.7
8	Shiny	West	Far/Shaded	16.0

data to see what, if any, factors prove to be significant. Make a recommendation on how to shade the car and how and where to park it. (Suggestion: use the software provided with this book. Set up a factorial design, similar to the one you did for the tutorial that comes with the program, for three factors in eight runs. Sort the design by standard order to match the table above and enter the data. Then do the analysis as outlined in the tutorial.)

CHAPTER 4
DEALING WITH NON-NORMALITY VIA RESPONSE TRANSFORMATIONS

"No experiments are useless."
—Thomas Edison

At the end of Chapter 3, we showed you how to check for normality—a fundamental assumption of the statistical analysis for DOE. In this chapter we discuss how to deal with non-normality (and nonconstant variance) via transformation of the response data. The most common transformation, the logarithm, is illustrated with a case study. This is the biggest DOE illustrated thus far: a two-level design on four factors, requiring 16 runs for all the combinations. After detailing this DOE, we will use fractional designs to squeeze more factors into the same number of runs.

Skating on Thin Ice

The data in this chapter comes from an exercise called "tabletop hockey." It works well as an in-class experiment for workshops on DOE. The objective of the hockey experiment is to learn how to shoot a puck for distance with a flexible, 15-centimeter (cm) ruler. The puck is comprised of two or more quarters (25-cent coins) stuck together with a gum adhesive. A simple wooden block acts as a fixture for the ruler. The response is the distance the puck slides over a smooth tabletop.

> ### WENT TO A FIGHT AND A HOCKEY GAME BROKE OUT!
> True hockey fans, particularly at the college level, appreciate the planning and organization of a well-coached team. The rink-long breakaways and resounding checks may arouse the crowd, but good passing and discipline win out in the end. Similarly, a hit-or-miss approach to experimentation may achieve goals in spectacular fashion, but the odds over the long haul favor the well-planned factorial approach.

After some brainstorming, a team of workshop students decided to study four factors at the levels shown in Table 4-1.

Figure 4-1 is a template for the tabletop hockey exercise. The circles show the various locations for the factor named "stick length." The marks below the line of the ruler relate to the factor named "windup." The last factor, named "puck place," is not shown on the figure. The low level for puck place is 0%, which means it is kept at the original position in front of the ruler and slapped. The high level for puck place is 100%, which puts it against the ruler so it can be flung forward, much like a wrist shot in the real game of hockey.

Table 4-2 lists the experiments in standard order with factors in coded form. The actual experiment was performed in random order. The response (Y) covers a broad range, from 3 to 190 centimeters—more than a 50-fold change. When data spans an order of magnitude (10-fold) or more, model fitting is often simplified by applying a logarithm to the response. Therefore, we added a column for the response in log base 10 form. This is called a "transformation."

Ultimately, we will show that the logarithmic scale works best, but for comparison sake, let's first analyze the data without a transformation. (In Table 4-2

Table 4-1. Test factors for table-top hockey

Factor	Name	Units	Low Level (−)	High Level (+)
A	Puck weight	Quarters	4	6
B	Stick Length	Centimeters	7.5	15
C	Windup	Centimeters	5	10
D	Puck place	Percent	0	100

DEALING WITH NON-NORMALITY VIA RESPONSE TRANSFORMATIONS

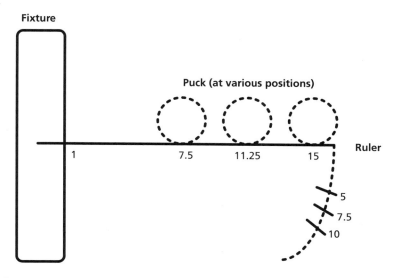

Figure 4-1. Template for tabletop hockey (dimensions in centimeters)

Table 4-2. Data for tabletop hockey (response Y is distance in centimeters)

Standard	A	B	C	D	Y	$\text{Log}_{10}Y$
1	−	−	−	−	38.2	1.582
2	+	−	−	−	23.3	1.367
3	−	+	−	−	3.0	0.477
4	+	+	−	−	7.6	0.881
5	−	−	+	−	110.0	2.041
6	+	−	+	−	90.6	1.957
7	−	+	+	−	20.6	1.314
8	+	+	+	−	18.9	1.276
9	−	−	−	+	36.6	1.563
10	+	−	−	+	38.0	1.580
11	−	+	−	+	47.4	1.658
12	+	+	−	+	44.9	1.652
13	−	−	+	+	190.0	2.279
14	+	−	+	+	116.8	2.067
15	−	+	+	+	137.5	2.138
16	+	+	+	+	84.5	1.927

Table 4-3. Table of effects

Term	Effect (Y)	Effect (Log Y)
A	-19.8375	-0.045
B	-34.8875	-0.39
C	66.2375	0.53
D	47.9375	0.50
AB	6.6875	0.078
AC	-16.9875	-0.091
AD	-11.9875	-0.062
BC	-26.5875	-0.035
BD	18.1125	0.36
CD	24.2375	-0.043
ABC	2.7875	-0.066
ACD	-14.2875	-0.088
ABD	-2.6125	-0.013
BCD	0.9625	-0.081
ABCD	3.2375	0.076

and those following, ignore the column(s) labeled "Log_{10}" for now.) From the 16 unique combinations that result from the 2^4 factorial ($2 \times 2 \times 2 \times 2 = 16$), 15 effects can be estimated, as shown in Table 4-3. We used a computer to do the calculations. It does the work much faster and more accurately than the normal human, excepting statisticians. The tricky part for nonstatisticians is understanding the terminology and interpreting the outputs. After going over all the nuts and bolts of two-level factorials in Chapter 3, you have nothing to fear.

Once again you see columns of data that are transformed by the logarithm. We will be getting back to these very soon, but for now let's proceed with the analysis of the data as-is. The next step is to view the untransformed effects (see Figure 4-2).

The students could not see any obvious division of effects from big to little, so they just picked the largest effect, C. The ANOVA (not shown) did show a significant probability for the resulting model, but the residual plots looked odd (see Figure 4-3).

DEALING WITH NON-NORMALITY VIA RESPONSE TRANSFORMATIONS

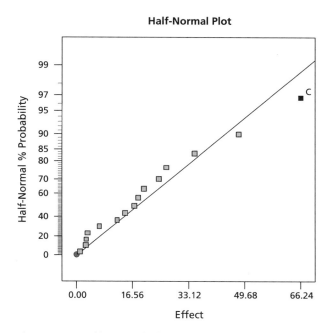

Figure 4-2. Half-normal plot of effects for tabletop hockey

The residuals did not line up on the normal plot, and they clearly increased with predicted level, forming the unwanted megaphone pattern. Something was wrong with the basic statistical assumptions.

Log Transformation Saves the Data

The tabletop hockey data exhibits a very common characteristic—as the response increases, so does the variance. This is a case of constant *percent* error. For example, a system may exhibit 10% error, so at a level of 10 the standard deviation is 1, but at a level of 100 the standard deviation becomes 10, and so on. Transforming such a response with the logarithm will stabilize the variance.

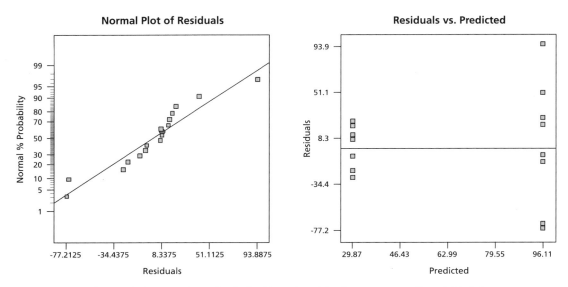

Figure 4-3 (a,b). Normal plot of residuals (left) and residuals versus predicted (right)

> ### *WHEN RESIDUALS MISBEHAVE, HIT THEM WITH A LOG*
> Logarithms are used as a scaling function for many measurements; for example, decibels of sound, Richter scale for earthquakes, pH rating of acidity, and astronomical units for stellar brightness. Try rescaling your response to log when residual diagnostic plots show abnormalities. However, don't expect much of an impact if the range of response is three-fold or less. In this case, the response transformation will be more trouble than it is worth. Also, remember that you cannot take the log of a negative number. This can be overcome by adding a constant to all the responses so that all become positive.

Luckily, we anticipated the need for a log after seeing the wide range of response. (Imagine that!) Using the transformed effects from Table 4-3, we generated the half-normal plot shown in Figure 4-4.

The transformation is amazing! It reveals a subset of relatively large effects, including interaction BD. As you should expect from seeing such a dramatic half-normal plot, analysis of variance indicates that all four chosen effects (B, C, D, and BD) are highly significant (see Table 4-4).

DEALING WITH NON-NORMALITY VIA RESPONSE TRANSFORMATIONS

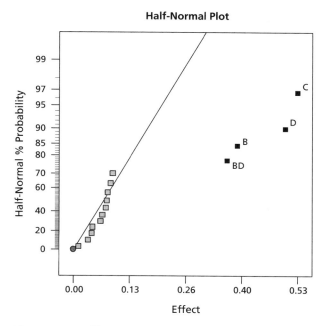

Figure 4-4. Half-normal plot of transformed effects (Log$_{10}$)

The residuals from the transformed model now look much better (see Figure 4-5).

In the final analysis, the distance of the shot is most affected by factor C: the windup of the ruler (see Figure 4-6).

Table 4-4. ANOVA for transformed response

Source	Sum of Squares	Df	Mean Square	F Value	Prob >F
Model	3.23	4	0.81	46.19	<0.0001
B	0.60	1	0.60	34.24	0.0001
C	1.11	1	1.11	63.66	<0.0001
D	0.99	1	0.99	56.76	<0.0001
BD	0.53	1	0.53	30.11	0.0002
Residual	0.19	11	0.017		
Cor Total	3.43	15			

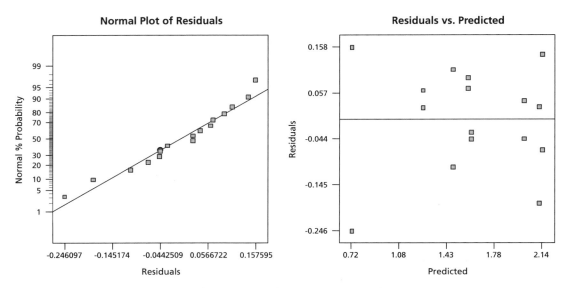

Figure 4-5 (a,b). Residuals plots after log transformation

As you would expect, larger windups produced longer shots. The weight of the puck (factor A) had no effect, at least over the range tested, so you can save two quarters!

The significant interaction between B and D is shown in Figure 4-7. The results may not be what you anticipated.

The effect of stick length (B) depends on puck placement (D). The D+ line is flat, with overlapping LSD bars at either end. This indicates that stick length makes no difference when you fling the puck from the stick (100% setting). However, when the puck is left at the original ruler line and slapped (D–), the longest shot comes with the shorter stick (B–). This is counterintuitive to most students who do this exercise. That's why it often pays to do an experiment, rather than relying on intuition.

To see the combined effect of the three significant factors, view the cube plots in Figure 4-8 (see p. 83). The left one is in transformed units to be consistent with the previous effect graphs done in the as-analyzed metric. But you really want to know the distance in original units. This can be generated by taking the antilog of the predicted responses. The right cube in Figure 4-8 presents the results after taking this reverse transformation.

DEALING WITH NON-NORMALITY VIA RESPONSE TRANSFORMATIONS

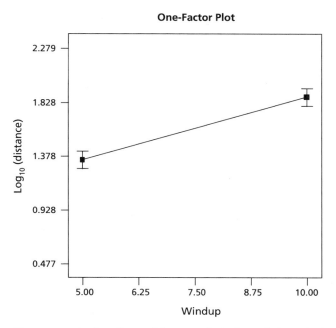

Figure 4-6. Main effect of factor C (windup of the ruler)

The best results—well over 100 centimeters—can be seen at the upper back side, with long windup (C+) and the puck placed against the stick (D+). At these settings of C and D, the stick length makes little difference, but by choosing the shorter level (B–) you lessen the impact of potential variations in the placement of the puck.

Choosing the Right Transformation

The abnormal residual plots shown on Figure 4-3 exhibit a not uncommon "power law" relationship between the standard deviation and the mean response. Statistically, this situation is symbolized as follows:

$$\sigma_y \propto \mu^\alpha$$

where the Greek letter sigma (σ) is the true standard deviation (of response Y), which is proportional to the true mean (mu-μ) to some power (alpha-α). Table 4-5 shows just a few of the possibilities for this power law, along with the appropriate transformations.

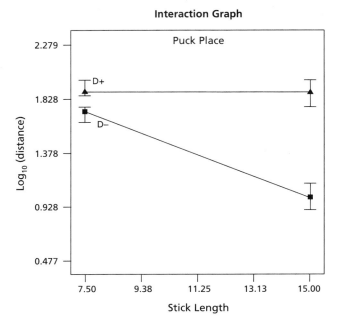

Figure 4-7. Interaction BD (stick length and puck place)

Ideally, there is no power law relationship ($\alpha = 0$), so no transformation is needed. In some cases, such as counts of imperfections, the standard deviation increases with the mean to the 0.5 power. The direct power relation ($\alpha = 1$) implies that the error is a constant percent of the response. This is a very common problem, which is remedied by a log transformation. In some cases, particularly when the response is a rate (for example, liters per second), the standard deviation increases with the square of the mean (power of 2). Then the inverse transformation is indicated (for example, seconds per liter).

Table 4-5. Variance-stabilizing transformations

Power (α)	Transformation	Comment
0	None	Normal
0.5	Square root	Counts
1	Logarithm	Constant percent error
2	Inverse	Rate data

DEALING WITH NON-NORMALITY VIA RESPONSE TRANSFORMATIONS

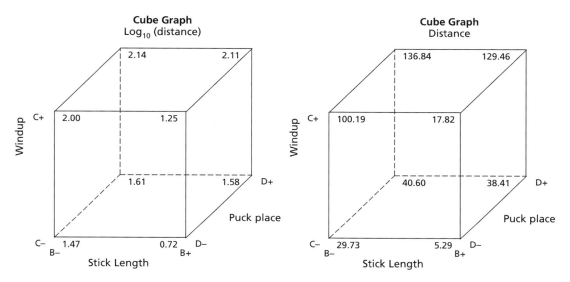

Figure 4-8 (a,b). Cube plot of predicted response in log (left) versus original units

---

THE DEATHLY COUNT

Counts of traffic accidents and deaths follow a distribution called "Poisson" where the standard deviation is a function of the mean. This was demonstrated by (Count?) Von Bortkiewicz, who kept track of Prussian cavalry soldiers killed by their horses between 1875 and 1894 ("Law of Small Numbers," 1898). Pity the poor soldiers who became statistics.

Transformations such as those shown above may stabilize the residual variance, satisfying the assumptions for ANOVA. They may be supported by scientific knowledge of the underlying relationship between factor(s) and response. For example, a Swedish chemist by the name of Arrhenius showed that temperature has an exponential effect on reaction rate. Typically, for every 10 degrees Celsius increase in temperature, reaction rate responds by quadrupling. In this case, a log transformation is the obvious remedy for linear modeling.

If you're uncertain whether a transformation would help, try one. However, if you don't see a definite improvement in the ANOVA (F-test) and residuals, don't bother with the transformation because it complicates matters. If you do use a transformation, remember to reverse the process by applying the reverse function, such as the antilog for a logged response. Otherwise you might get some grief about your goofy predictions!

Practice Problem

PROBLEM 4-1

This problem demonstrates that design of experiments can be applied to any system, even one that does not involve manufacturing. It addresses a question debated by producers of goods and services aimed at a technical audience: Would this data-driven personality type react favorably to fancy four-color printing on a direct-mail piece? Conventional wisdom says the answer will be yes, but maybe this applies only to nontechnical consumers. Perhaps cheaper two-color printing would work as well, or better, for technical types.

DO WE NEED ANOTHER ACRONYM FOR DESIGN OF EXPERIMENTS?

Forbes magazine, in the March 11, 1996 issue, introduced the concept of design of experiments to the business world. The title of the article, "The New Mantra MVT," introduced a new acronym (MVT) that stands for multivariable testing. Admittedly, the reference to "multivariable testing" conveys a major benefit to this style of experimentation. However, the case studies show that MVT is simply design of experiments applied to business and marketing problems. Whatever you call it, business people should take notice and take advantage of these powerful methods.

"If you test factors one at a time, there's a very low probability that you're going to hit the right one before everybody gets sick of it and quits."

—Forbes magazine

In addition to the color factor, market researchers looked at two postcard sizes (small versus big) and two types of paper stock (thin versus thick). The eight resulting postcard designs (2^3) were sent to eight equal segments of the company's client list, chosen at random. To garner more response, the researcher offered a free technical report to anyone who faxed back the reply side of the postcard. The postcards incorporated the standard two-level code to facilitate

DEALING WITH NON-NORMALITY VIA RESPONSE TRANSFORMATIONS

Table 4-6. Results from postcard experiment

Standard	A: Color	B: Size	C: Thickness	Requests (count)	Printing Cost (cents/card)
1	Two	Small	Thin	152	6
2	Four	Small	Thin	57	10
3	Two	Large	Thin	248	8
4	Four	Large	Thin	31	12
5	Two	Small	Thick	250	8
6	Four	Small	Thick	131	12
7	Two	Large	Thick	398	10
8	Four	Large	Thick	96	14

measurement. For example, the first and last combinations in standard (Std) order were coded:

- − − − (= two-color, small card on thin stock)
- + + + (= four-color, big card on thick stock)

Table 4-6 shows the number of requests generated by each postcard configuration.

The cost of printing the cards is also shown for reference. This is a "deterministic" response because it depends only on the factor levels. The four-color, large, thick postcard was the most expensive combination.

> ### OBTAIN ENOUGH RESPONSES TO GENERATE STATISTICAL SIGNIFICANCE
> To get good, reliable results from tests on direct-mail pieces, you must generate a significant response from every configuration. Market researchers advise a minimum of 20 responses per row, or "cell" as they call it. For simple comparison, where only two varieties are tested on a split mail-list, the following rule-of-thumb is often applied: If the difference between the test results is two-times greater than the square root of the total, it is a significant difference. Because responses may continue to trickle in for many weeks, market researchers often extrapolate the early returns in order to generate preliminary findings. For example, they might double the response received after two weeks time.

Analyze this data. Given that this section of the book focuses on use of transformations, consider trying one. (Hint: the response is a count.) Determine the combination that maximizes response. You might be surprised by the results! (Suggestion: use the software provided with this book. Set up a factorial design, similar to the one you did for the tutorial that comes with the program, for three factors in eight runs with two responses. Sort the design by standard order to match the table above, enter the data, and do the analysis as outlined in the tutorial. Then go back and re-analyze after first choosing the square root as a response transformation. Compare the model and resulting residual plots before and after doing the transformation.)

CHAPTER 5
FRACTIONAL FACTORIALS

"Believe nothing and be on your guard against everything."
—Latin Proverb

The full-factorial approach to experimentation covers all combinations of factors, providing valuable information on interactions. However, the number of experimental runs increases rapidly, even if you test the factors at only two levels each. Fortunately, by resorting to a "fractional factorial," you can study many factors and still keep the experiment to a reasonable size. Table 5-1 shows the potential savings for four to seven factors (see Appendix 2 for details on layout). These particular designs are especially good for "screening" many factors in search of the vital few. For reference, we show the number of main effects and two-factor interactions. An efficient fraction contains at least this many runs plus one more for estimating the overall average, but no more. The five-factor, 16-run fraction is an ideal example. For five or more factors, full factorials are not efficient.

We hope this whets your appetite, because we will soon get into the nitty-gritty of fractional factorial design, including a more complete table than that shown below. You will learn that there's a price to pay in the form of "aliasing"

Table 5-1. Selected two-level fractional factorial screening designs

Factors	Main Effects and Two-Factor Interactions	Full Factorial	Fraction for Screening	Savings
4	10	16	12	25%
5	15	32	16	50%
6	21	64	32	50%
7	28	128	32	75%

effects. The more you know about this the better, because conducting fractional factorials is like playing with fire. It's a powerful tool that could burn you if not handled carefully.

Example of Fractional Factorial at Its Finest

Let's begin with an example of a very safe fractional factorial: five factors at two levels each, done in 16 runs (a half-fraction). This data comes from an actual experiment on an uncooperative grass trimmer, commonly known in the Midwest as a weed whacker. A one-cylinder engine that runs on a mixture of gasoline and oil powers the weed whacker (Figure 5-1). Perhaps you have become frustrated trying to start small engines such as this one. It features a manual starter (pull cord), and its performance is largely a function of three controls: primer pump, choke, and gas. Before starting the engine, you must first give the primer several pumps, choke the air intake, and pull the starter-cord several times. To start the engine, the choke must be open. Table 5-2 lists the tested factors and levels.

It would require 32 runs to perform all the combinations of these five factors. This full, two-level factorial would reveal 31 effects: 5 main effects, 10 two-factor interactions, 10 three-factor interactions, 5 four-factor interactions,

Figure 5-1. The weed-whacker engine

Table 5-2. Factors for weed-whacker experiment

Factor	Name	Low Level (–)	High Level (+)
A	Prime pumps	3	5
B	Pulls at full choke	3	5
C	Gas during full choke pulls	0	100%
D	Final choke setting	0	50%
E	Gas for start	0	100%

and one five-factor interaction. Interactions of three or more factors are highly unlikely in an engine or any other system. Moreover, the principle of sparsity of effects states that only 20% of the main effects and two-factor interactions are likely to be significant in any particular system. If this is true, then only three effects will be significant—which leaves 28 effects for

Table 5-3. Design layout for weed-whacker experiment

Std	A	B	C	D	E	Pulls
1	–	–	–	–	+	1
2	+	–	–	–	–	4
3	–	+	–	–	–	4
4	+	+	–	–	+	2
5	–	–	+	–	–	8
6	+	–	+	–	+	2
7	–	+	+	–	+	3
8	+	+	+	–	–	5
9	–	–	–	+	–	3
10	+	–	–	+	+	1
11	–	+	–	+	+	3
12	+	+	–	+	–	4
13	–	–	+	+	+	3
14	+	–	+	+	–	4
15	–	+	+	+	–	6
16	+	+	+	+	+	5

estimation of error, far more than necessary. Therefore, a full factorial on five factors (or more) will waste much of its effects on unneeded estimates of error.

A properly constructed half-fraction (16 runs) estimates the overall mean, the five main effects, and the 10 two-factor interactions. The design for the weed whacker, constructed by standard methods, is shown in Table 5-3. Experiments are listed in standard order, not randomized run order. Factors are shown in coded format. The response (Y) is pulls needed to start the engine.

You can use this as a template for your own experiment on five factors. Notice that the first four factor columns form a full factorial with 16 runs. The fifth factor column (E) is the product of the first four columns (ABCD). Check it out! This and many other fractional designs can be generated via statistical software, or from tables in referenced textbooks on DOE. We will give you more clues on their construction later on in this section.

The half-normal plot of effects (Figure 5-2) reveals two large main effects: factors E and C. Be careful, though; remember that E was constructed from

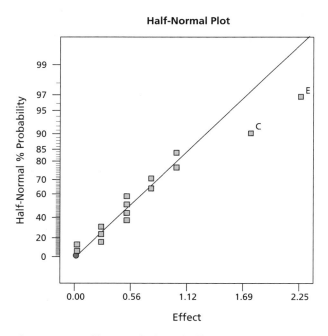

Figure 5-2. Half-normal plot of effects for weed-whacker

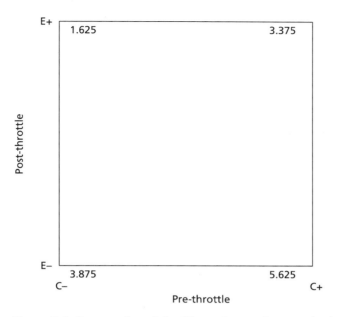

Figure 5-3. Square plot of significant factors for weed-whacker

ABCD. These two effects are "aliased." The effect of C is also aliased with a four-factor interaction: ABDE. (Check this out by multiplying the appropriate columns and comparing signs.) Four-factor interactions are not plausible, so we will ignore these aliases. This assumption follows the generally accepted practice of statisticians. However, you should never forget that you can't get something for nothing—fractional factorials save on runs, but they produce aliases, which can be troublesome.

The F-test on the resulting model was significant at a 99.9% confidence level. Residual analysis showed no abnormalities. Figure 5-3 shows the four combinations of the two significant factors. The best combination (least number of pulls) occurs at low C and high E at the upper left of the square plot.

> ### *POWER TO CUT THE GRASS*
> Fractional factorial designs, such as those shown in this book, exhibit a property called "projection." This means that the subset of significant factors becomes equivalent to a full factorial design. For example, in the weed-whacker experiment, the design projects to a two-by-two factorial replicated four times. Each point on the square plot in Figure 5-3 is based on an average of four results. This provides greater power to see small effects hidden in the underlying variation, much like an engine-powered weed whacker cuts through tall grass to reveal lost golf balls and the like.

None of the other factors matter, so they can be set at their most advantageous level. The ideal procedure for minimizing pulls is:

- Prepare engine with three primer pumps and three pulls at 100% choke with 0% gas
- Start at 0% choke at 100% gas.

In the confirmation trial, the engine started immediately on the first pull!

Potential Confusion Caused by Aliasing in Lower-Resolution Factorials

Does the application of fractions to factorial design appear to be too good to be true? Obviously, you don't get something for nothing. The price you pay when you cut down the number of runs is the aliasing of effects. This is measured by the "resolution" of the fractional factorial. The half-fractional design on the five weed-whacker factors is a Resolution V (the resolution is represented by Roman numerals), which indicates aliasing of at least one main effect(s) with one or more four-factor interaction(s), and/or at least one two-factor interaction(s) with one or more three-factor interaction(s). Think of these as relationships of 1 with 4 and 2 with 3, both of which total to 5. You can figure this out with your fingers!

As you might expect, as resolution decreases, alias problems increase. Let's go to a worst-case scenario for two-level factorials—a Resolution III design. We will use the popcorn case from Chapter 3 as a demonstration. The unshaded

FRACTIONAL FACTORIALS

Table 5-4: Popcorn experiment with half of runs (shaded) removed

Std.	A	B	C	AB	AC	BC	ABC	Taste
1	−	−	−	+	+	+	−	74
2	+	−	−	−	−	+	+	75
3	−	+	−	−	+	−	+	71
4	+	+	−	+	−	−	−	80
5	−	−	+	+	−	−	+	81
6	+	−	+	−	+	−	−	77
7	−	+	+	−	−	+	−	42
8	+	+	+	+	+	+	+	32
Effect	-1.0	-20.5	-17.0	0.5	-6.0	-21.5	-3.5	Full
Effect	-22.5	-26.5	-16.5	-16.5	-26.5	-22.5	64.75	Fraction

rows in Table 5-4 represent a half fraction of the original data. It was created by eliminating the minus ABC runs (shaded in the table).

Can you see any problems with the resulting pattern of pluses and minuses in the clear rows?

At first glance, the design looks good, with a nicely balanced pattern for the main effects of A, B, and C. But upon closer inspection, notice that each effect column has an identical twin in terms of the pattern of pluses and minuses (shown below in parentheses) you see from top to bottom in the clear rows:

- A = BC (+, −, −, +)
- B = AC (−, +, −, +)
- C = AB (−, −, +, +)

These equalities are called "confounding" relationships, or aliases. The most troublesome of these aliases involves the interaction BC. Recall that the full factorial revealed this to be a significant effect. However, as shown above, the half-fraction attributes the BC effect to A, which is completely misleading. The effect labeled A is actually the combination of A and BC, which can be expressed mathematically as: A = A + BC. (To check this equation, observe that the effects of A and BC from the full factorial are −1.0 and −21.5, respectively, which sum to the value of −22.5 for A in the half-fraction.)

Finally, notice that the last effect column, ABC, displays only plus symbols in the clear rows of Table 5-4. Since there is no contrast in the signs, this effect can no longer be estimated. The calculated value of 64.75, labeled as an effect, is actually an estimate of the overall mean or intercept.

You can now see the downside to creating a fractional design: aliasing of effects. In this case, we cut a full factorial to a half-fraction by eliminating the negative ABC rows. The resulting loss of information about the ABC effect is not critical, because three-factor interactions rarely occur. However, you should be very concerned about the aliasing of main effects with two-factor interactions. Statisticians consider such a design to be Resolution III, the lowest possible for standard fractional two-level factorials. We recommend that you avoid Resolution III designs if at all possible, because if your system contains any interactions, the true cause will be disguised by the aliasing. However, in anticipation that some of you will ignore this advice, we will provide more details on Resolution III designs and show how to de-alias them in the next chapter.

DESPERATE MEASURES

DOE guru George Box says that Resolution III designs are like kicking the television to make it work. You may succeed but you won't know which component dropped into place or whatever else actually caused the improvement.

"[Don't] use statistics as a drunken man uses lamp posts, for support rather than illumination."

—Andrew Lang

So far, we've shown examples of Resolution V (good) and Resolution III (bad). Let's fill the gap with a Resolution IV design, which represents a reasonable compromise between minimal runs and maximum information on the main effects. Table 5-5 shows a complete matrix of effects for four factors. The rows where ABCD is minus have been grayed out. The remaining rows form the half-fraction. Look over the resulting columns very carefully, ignoring the gray cells. Can you see the aliases? (Hint: start in the middle, at AD–BC and work outward.)

In this case, every main effect is aliased with a three-factor interaction, and all the two-factor interactions are aliased with each other, so this design is Resolution IV.

FRACTIONAL FACTORIALS

Table 5-5. Four-factor design matrix with half-fraction unshaded

Std	A	B	C	D	AB	AC	AD	BC	BD	CD	ABC	ABD	ACD	BCD	ABCD
1	−	−	−	−	+	+	+	+	+	+	−	−	−	−	+
2	+	−	−	−	−	−	−	+	+	+	+	+	+	−	−
3	−	+	−	−	−	+	+	−	−	+	+	+	−	+	−
4	+	+	−	−	+	−	−	−	−	+	−	−	+	+	+
5	−	−	+	−	+	−	+	−	+	−	+	−	+	+	−
6	+	−	+	−	−	+	−	−	+	−	−	+	−	+	+
7	−	+	+	−	−	−	+	+	−	−	−	+	+	−	+
8	+	+	+	−	+	+	−	+	−	−	+	−	−	−	−
9	−	−	−	+	+	+	−	+	−	−	−	+	+	+	−
10	+	−	−	+	−	−	+	+	−	−	+	−	−	+	+
11	−	+	−	+	−	+	−	−	+	−	+	−	+	−	+
12	+	+	−	+	+	−	+	−	+	−	−	+	−	−	−
13	−	−	+	+	+	−	−	−	−	+	+	+	−	−	+
14	+	−	+	+	−	+	+	−	−	+	−	−	+	−	−
15	−	+	+	+	−	−	−	+	+	+	−	−	−	+	−
16	+	+	+	+	+	+	+	+	+	+	+	+	+	+	+

A HANDY WAY TO PUT YOUR FINGER ON THE CONCEPT OF RESOLUTION

To determine the alias structure of a given resolution, count it with your fingers. For example, in a Resolution III design, hold up three fingers. In this design, at least one main effect (represented by the first finger or thumb) is aliased with at least one two-factor interaction (represented by the other two fingers). It's very simple—one plus two equals three. If you choose a Resolution IV design, hold up four fingers. In this design, at least one main effect (represented by the first finger) is aliased with at least one three-factor interaction (represented by the other three fingers). One plus three equals four. Another option presents itself in this case: two plus two also equals four. This represents the presence of at least one alias between a pair of two-factor interactions. If you don't like this, choose a Resolution V design. You now need a hand full of fingers, because main effects are aliased only with four-factor interactions (thumb plus four fingers), and two-factor interactions are aliased only with three-factor interactions (two plus three equals five).

Assuming that three-factor interactions are unlikely, you can rely on main effect estimates from Resolution IV designs, but don't jump to any conclusions on any significant two-factor interactions. For example, if AB shows significance, it might really be due to CD, which changes in exactly the same way: +, +, −, −, −, −, +, + from top to bottom, clear rows only. We will show you how to escape from this trap in the next chapter of the book.

> ### BLOCKING TWO-LEVEL FACTORIALS
> What if you had only enough time in a day to do half the runs in a full four-factor design? You could run the half fraction shown in Table 5-5, but perhaps you don't want to take a chance on the resulting aliasing of two-factor interactions. In this case, you could run the first fraction on day one and the second fraction on day two. Within each day you should randomize the run order. Each day must be considered as a block of time. The block difference can be removed in the analysis, but only at a cost—the estimate of ABCD will be sacrificed. This may be a small price to pay in return for gaining resolution of the two-factor (and three-factor) interactions. Using similar approaches to those used for generating fractional designs, statisticians have worked out optimal schemes to block two-level factorials into two, four, or more groups. The objective is to sacrifice a minimal number of lower-order effects. As discussed earlier in this book, the tool of blocking is very powerful for removing known sources of variation, such as time or material or operators.

Not all Resolution IV designs are the same: some have only a few two-factor interactions aliased with each other. For example, the seven factors in the 32-run design (1/4th fraction) shown on Table 5-6 is listed as a Resolution IV, but only six of the two-factor interactions (those involving D, E, F, and G) are actually aliased (DE = FG, DF = EG, and DG = EF). Other fifteen two-factor interactions are aliased only with three-factor or higher interactions. Therefore, you would be wise to assign factors that are least likely to interact with the labels D, E, F, or G; and those factors most likely to interact with A, B, and C. If you want details on the specific layout for this design (seven factors in 32 fraction), see Appendix 2-4.

FRACTIONAL FACTORIALS

Table 5-6. Resolutions for standard two-level designs with reasonable number of runs

Factors	2	3	4	5	6	7	8
4 Runs	Full	1/2 III	—	—	—	—	—
8	2 Rep	Full	1/2 IV	1/4 III	1/8 III	1/16 III	—
16	4 Rep	2 Rep	Full	1/2 V	1/4 IV	1/8 IV	1/16 IV
32	8 Rep	4 Rep	2 Rep	Full	1/2 VI	1/4 IV	1/8 IV
64	16 Rep	8 Rep	4 Rep	2 Rep	Full	1/2 VII	1/4 V

On the other hand, you might be tempted to cut back to only 16 runs for the seven-factor design, a 1/8th fraction. After all, as shown in Table 5-6, it too is Resolution IV. However, as you can see from the alias structure shown in Table 5-7, no matter how you do the labeling, even one active interaction may cause confusion. For example, let's say that only C and D interact in a particular system. This interaction (CD) will be mislabeled as AG due to the aliasing.

It should be apparent by now that you must investigate the alias structure for anything less than Resolution V designs. You will find the details on construction of optimal fractions and their alias structures in the referenced textbooks by Box and Montgomery. Good DOE software will also set up the appropriate fraction and supply data on the specific aliases of your chosen design.

Table 5-7. Alias structure for seven factors in 16 runs (two-factor interactions only)

Labeled as	Actually
AB	AB + CE + FG
AC	AC + BE + DG
AD	AD + CG + EF
AE	AE + BC + DF
AF	AF + BG + DE
AG	AG + BF + CD
BD	BD + CF + EG

Plackett-Burman Designs

The standard two-level designs, which we recommend, provide the choice of 4, 8, 16, 32, or more runs, but only to the power of two. In 1946, Plackett and Burman invented alternative two-level designs that are multiples of four. The 12-, 20-, 24-, and 28-run Plackett-Burman designs are of particular interest because they fill gaps in the standard designs. Unfortunately, these particular Plackett-Burman designs have very messy alias structures. For example, the 11 factor in the 12-run choice, which is very popular, causes each main effect to be partially aliased with 45 two-factor interactions. In theory, you can get away with this if absolutely no interactions are present, but this is a very dangerous assumption in our opinion.

Of course, you could cut down the number of factors in a given Plackett-Burman, but the alias structure may not be optimal. Assuming that your software provides such data, be sure to look at the alias structure before doing your experiment. Because of the unexpected aliasing that occurs with many Plackett-Burman designs, we recommend that you avoid them in favor of the standard two-level designs. However, if you go ahead with a Plackett-Burman, consider following up with a second set of runs, but not an exact replicate. In the next chapter, we will show you how to do a "fold over" that doubles the number of runs in a way that increases the resolution of highly aliased Plackett-Burman designs and standard fractional factorials.

Irregular Fractions Provide a Clearer View

We've shown options for the standard two-level design which apply fractions that are negative powers of 2 (1/2, 1/4, etc.). However, it is possible to do other "irregular" fractions and still maintain a relatively high resolution. A prime example of this is the three-quarter (3/4) replication ("rep") for four factors (the first design in Table 5-1). It can be created by identifying the standard quarter-fraction, and then selecting two more quarter-fractions. This design contains only 12 runs, yet it estimates all main effects and two-factor interactions aliased only by three-factor or higher interactions, thus making it a viable alternative to the 16-run full factorial. (Details on this and other designs of the like can be found in the referenced text by John.) We will now work through an example of the design with four factors in 12 runs.

FRACTIONAL FACTORIALS

Table 5-8. Test factors for RGB projector study

Factor	Name	Units	Low Level (−)	High Level (+)
A	Font size	Point	10	18
B	Font style	(Categorical)	Arial	Times
C	Background	(Categorical)	Black	White
D	Lighting	(Categorical)	Off	On

The authors frequently present computer-intensive workshops on DOE in corporate training rooms equipped with RGB (red-green-blue) projectors. Students often find it difficult to read the projected statistical outputs. To improve readability, we investigated four factors at two levels, as shown in Table 5-8. The assignment of minus versus plus levels for the categorical variables (B, C, D) is completely arbitrary.

Students worked together on the exercise. The instructor displayed a column of numbers in random order. One student transcribed the data from top to bottom while the other timed it. The actual experiment was performed by

Table 5-9. Irregular fraction design on RGB projection

Std	A: Font Size (point)	B: Font Style	C: Background	D: Lighting	Y: Readability (seconds)
1	10 (−)	Arial (−)	Black (−)	Off (−)	52
2	18 (+)	Times (+)	Black (−)	Off (−)	39
3	10 (−)	Arial (−)	White (+)	Off (−)	42
4	18 (+)	Arial (−)	White (+)	Off (−)	27
5	10 (−)	Times (+)	White (+)	Off (−)	37
6	18 (+)	Times (+)	White (+)	Off (−)	31
7	10 (−)	Arial (−)	Black (−)	On (+)	57
8	18 (+)	Arial (−)	Black (−)	On (+)	28
9	10 (−)	Times (+)	Black (−)	On (+)	52
10	18 (+)	Times (+)	Black (−)	On (+)	30
11	18 (+)	Arial (−)	White (+)	On (+)	19
12	10 (−)	Times (+)	White (+)	On (+)	47

many students. By treating each student as an individual block, we eliminated variability caused by differing distance from the screen and reading ability. However, to keep it simple, the results are shown as one block of data. Table 5-9 provides the complete design with the resulting readability measured in seconds. Shorter times are desired.

The four main effects (A, B, C, D) plus six two-factor interactions (AB, AC, AD, BC, BD, CD) are calculated in the usual way by contrasting the average of the plus values with average of the minus values for the associated response (symbolized by Y). For example, the main effect of factor A is:

$$\text{Effect}_A = \frac{\sum Y_{A+}}{n_{A+}} - \frac{\sum Y_{A-}}{n_{A-}} = \frac{39 + 27 + 31 + 28 + 30 + 19}{6} - \frac{52 + 42 + 37 + 57 + 52 + 47}{6}$$

$$= 29.00 - 47.83 = -18.83$$

In other words, students read the projected display 18.83 seconds faster (on average) at a type size of 18 versus 10 points. In this case, the bigger the bet-

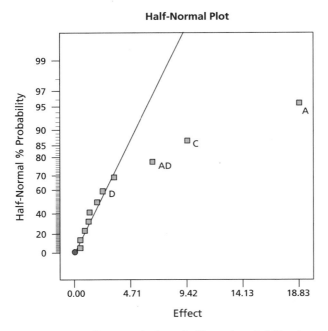

Figure 5-4. Half-normal plot of effects (readability in seconds) for projector DOE

ter! The effect of A (font size), in absolute value, stands out on the half-normal plot (Figure 5-4).

The main effects of C (background) and the interaction AD (font size times lighting) also fall off the normally distributed line of near-zero effects. As usual, we did not label any of the smaller, presumably insignificant effects, with one exception: D. For statistical reasons which will be discussed later, this main effect must be chosen to support the choice of the AD interaction.

> ### WARNING: IRREGULAR FRACTIONS MAY PRODUCE IRREGULARITIES IN EFFECT ESTIMATES
>
> Irregular fractions have somewhat peculiar alias structures. For example, when evaluated for fitting a two-factor interaction model, they exhibit good properties: main effects aliased with three-factor interaction, etc. But if you fit only the main effects, they become partially aliased with one or more two-factor interactions. Appendix 2-1 provides the details for the four-factor in 12-run design. Normally this peculiarity would not be a problem because you would never restrict yourself to main effects only. However, certain software programs, such as the one used to generate outputs for this book, calculate effects hierarchically, starting with main effects. This is done to make the effect plotting more robust to missing data and/or changed factor levels. Under these circumstances information will be lost, so you want to limit the damage by first calculating the main effects. Full factorials and regular fractions will not be affected by the hierarchical effect estimation, but main effects from irregular designs get inflated by any large 2-factor interactions. The bad news is that insignificant main effects might be selected as a result. The good news is that you can easily catch these bogus effects by looking at the probability values in the analysis of variance for the selected model terms. If there are no interactions, or they are relatively small (such as in the RGB case), you won't be confronted with this anomaly.

Table 5-10. Analysis of variance for RGB readability data

Source	Sum of Squares	Df	Mean Square	F Value	Prob >F
Model	1501.58	4	375.40	60.64	<0.0001
A	1064.08	1	1064.08	171.89	<0.0001
C	266.67	1	266.67	43.08	0.0003
D	16.67	1	16.67	2.69	0.1448
AD	168.75	1	168.75	27.26	0.0012
Residual	43.33	7	6.19		
Cor Total	1544.92	11			

Table 5-10 shows the analysis of variance for the RGB data.

You may still be curious as to why we include factor D in the model, because it's not significant in the ANOVA. The factor was chosen to support the significant AD interaction, thus maintaining model "hierarchy" (see related sidebar).

PRESERVING FAMILY UNITY

Statisticians advise that you maintain hierarchy in your regression models. The idea of hierarchy can be likened to the traditional structure of a family, with parents and children. In this analogy, a two-factor interaction such as AD is considered a child. To maintain family hierarchy, you must include both parents (A and D). Similarly, if you select a two-factor interaction for a predictive model, be sure to also select both of the main effects. Otherwise, under certain circumstances, some statistics may not get computed correctly. (For details, see J. L. Peixoto, "A Property of Well-Formulated Polynomial Regression Models," *The American Statistician*, Feb. 1990, V44, No. 1.) However, an even better reason to abide by this rule is to avoid misleading your audience by saying a factor is not significant when it really does make a difference, albeit only in conjunction with one or more other factors. The RGB experiment is a case in point.

Although factor D is not significant on its own, it does have an effect on readability, but only in conjunction with factor A. This becomes very clear when you look at the interaction graph in Figure 5-5.

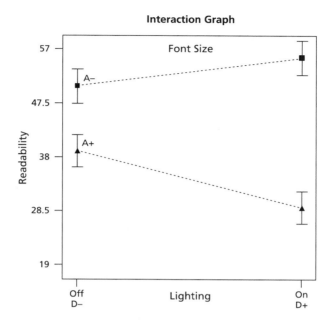

Figure 5-5. Interaction for factors A and D

If you split the differences from left (D–) to right (D+), you get a nearly flat line, indicating that D has no effect, thus explaining why it falls on the near-zero effect line shown in the half-normal plot. However, to say that D has no effect makes no sense. Factor D does affect the response, but it depends on the level of factor A (font size). When A is low (–), increasing D increases the response. But when A is high (+), increasing D *decreases* the response. In either case, factor D does cause an effect. Therefore, it should not be excluded.

From a practical perspective, the upper line on the AD interaction tells us that students find it more difficult to read the RGB screen when the font size is small (A–) with the lights turned on (D+). The lower line shows that when font size is large enough, it doesn't matter if you turn on the lights; in fact, the readability results actually improved. This is a very desirable outcome, because students can read their notes while viewing the RGB screen. Therefore, we concluded that the larger font size definitely is better.

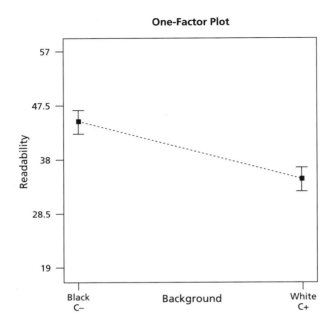

Figure 5-6. Main effect plot for factor C (background)

One other effect stood out on the half-normal plot: the main effect of C (background). As you can see in Figure 5-6, readability improves at the high level of this factor that represents background set to white.

Finally, we put all three of the significant factors together in the form of the cube plot shown in Figure 5-7. It shows the best result (19 seconds) at the right, back, upper corner where all factors are set at their high levels.

Note that B (font style) was not significant, so either one can be chosen. One author prefers Arial, and the other Times New Roman style, but we don't worry about it because it probably makes little or no difference in readability.

FRACTIONAL FACTORIALS

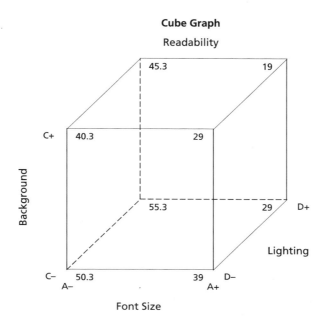

Figure 5-7. Cube plot of RGB readability as a function of factors A, C, and D

Practice Problem

PROBLEM 5-1

An injection molder wanted to gain better control of part shrinkage. The experimenter set aside two parallel production lines for a study of seven factors (see Table 5-11).

Table 5-11. Factors for molding case

	Factor	Units	Low (−)	High (+)
A	Mold temperature	degrees F	130	180
B	Cycle time	seconds	25	30
C	Booster pressure	psig	1500	1800
D	Moisture	percent	0.05	0.15
E	Screw speed	inches/sec	1.5	4.0
F	Holding pressure	psig	1200	1500
G	Gate size	inches (10^{-3})	30	50

All possible combinations of these factors require 128 experiments, but only 32 of these were actually run—a 1/4th fraction. This DOE is symbolized mathematically as 2^{7-2}. It's one of the recommended fractional factorial designs for screening (see Table 5-1, p. 87). From the 32 runs you can get information on all the main effects and nearly all two-factor interactions. Table 5-12 shows the design matrix in terms of coded factor levels, and the results for shrinkage.

Notice that the experiment is divided into two blocks of sixteen runs in a standard way that preserved the greatest possible information on main effects and interactions. The experimenters then ran the DOE on the two parallel lines, greatly reducing the time needed to generate the data, as well as providing information on machine-to-machine variation.

The blocking does cause additional loss of information in the form of additional aliasing (revealed by statistical DOE software):

- [Block 1] = Block 1 + CDG + CEF + ABDE + ABFG
- [Block 2] = Block 2 − CDG − CEF − ABDE − ABFG

The loss of the three-factor and higher interactions didn't cause much worry. They also carefully reviewed the alias structure (see Appendix 2-4) before assigning labels. By labeling the most likely interactors, booster pressure and moisture, as C and D, they avoided deliberate aliasing of this potential effect with other two-factor interactions.

Analyze the data. Look for conditions that minimize and/or stabilize shrinkage. Remember to check the significant factors against the alias structure in Appendix 2-4. (Suggestion: use the software provided with the book. Look for a data file called "Prb5-1-molding" and open it. Otherwise, create the design, by choosing a factorial for 7 factors in 32 runs with 2 blocks, and enter the data [from Table 5-12] in standard order. Then do the analysis as outlined in the factorial tutorial that comes with the program.)

FRACTIONAL FACTORIALS

Table 5-12. Design matrix for molding case study

Std	Run	Line	A	B	C	D	E	F	G	Shrinkage (%)
1	19	1	−1	−1	−1	−1	−1	+1	+1	0.833
2	4	2	+1	−1	−1	−1	−1	−1	−1	0.784
3	10	2	−1	+1	−1	−1	−1	−1	−1	0.966
4	25	1	+1	+1	−1	−1	−1	+1	+1	0.898
5	28	1	−1	−1	+1	−1	−1	−1	−1	0.916
6	11	2	+1	−1	+1	−1	−1	+1	+1	1.130
7	3	2	−1	+1	+1	−1	−1	+1	+1	0.760
8	29	1	+1	+1	+1	−1	−1	−1	−1	0.730
9	13	2	−1	−1	−1	+1	−1	−1	+1	0.838
10	17	1	+1	−1	−1	+1	−1	+1	−1	0.669
11	27	1	−1	+1	−1	+1	−1	+1	−1	1.060
12	14	2	+1	+1	−1	+1	−1	−1	+1	0.956
13	8	2	−1	−1	+1	+1	−1	+1	−1	1.780
14	22	1	+1	−1	+1	+1	−1	−1	+1	1.660
15	18	1	−1	+1	+1	+1	−1	−1	+1	1.080
16	12	2	+1	+1	+1	+1	−1	+1	−1	1.230
17	16	2	−1	−1	−1	−1	+1	+1	−1	0.922
18	24	1	+1	−1	−1	−1	+1	−1	+1	0.815
19	20	1	−1	+1	−1	−1	+1	−1	+1	1.100
20	9	2	+1	+1	−1	−1	+1	+1	−1	0.858
21	2	2	−1	−1	+1	−1	+1	−1	+1	1.170
22	30	1	+1	−1	+1	−1	+1	+1	−1	1.040
23	26	1	−1	+1	+1	−1	+1	+1	−1	0.780
24	15	2	+1	+1	+1	−1	+1	−1	+1	1.020
25	21	1	−1	−1	−1	+1	+1	−1	−1	0.939
26	1	2	+1	−1	−1	+1	+1	+1	+1	0.909
27	5	2	−1	+1	−1	+1	+1	+1	+1	1.060
28	31	1	+1	+1	−1	+1	+1	−1	−1	0.916
29	23	1	−1	−1	+1	+1	+1	+1	+1	1.680
30	6	2	+1	−1	+1	+1	+1	−1	−1	1.440
31	7	2	−1	+1	+1	+1	+1	−1	−1	1.330
32	32	1	+1	+1	+1	+1	+1	+1	+1	1.210

CHAPTER 6
GETTING THE MOST FROM MINIMAL-RUN DESIGNS

"The best carpenters make the fewest chips."
—GERMAN PROVERB

In the previous chapter, we demonstrated how to shave runs from a two-level factorial design by performing only a fraction of all possible combinations. In this chapter, we explore minimal designs with one fewer factor than the number of runs; for example, seven factors in eight runs. Statisticians consider such designs to be "saturated" with factors. These Resolution III designs confound main effects with two-factor interactions—a major weakness. However, they may be the best you can do when confronted with a lack of time or other resources. If you're lucky, nothing will be significant, because in that case any questions about aliasing become moot. However, if the results exhibit significance, you must make a big leap of faith to assume that the reported effects are correct. To be safe, do further experimentation—known as "design augmentation"—to de-alias the main effects and/or two-factor interactions. The most popular method of design augmentation is called the fold-over. We will illustrate this method via a case study.

> ### RUGGEDNESS TESTING
> Before transferring any system, such as a product, process, or test method, find out what the recipients will do to it. For example, a small U.S. medical device manufacturer redesigned the electronics in their 220-volt unit for European customers. It worked without fail in all but one country, where every single unit burned out. The country where this occurred was relatively undeveloped, and their power supply varied more than anyplace else in Europe. One of the components in the new design could not tolerate such wide variation in voltage. After this fiasco, the engineers developed a standard fractional-factorial design to test new designs against this variation and half a dozen or so other variables that units might encounter. This led to redesigning components and the overall system to make it more robust.

Minimal Resolution Design: The Dancing Raisin Experiment

The dancing raisin experiment provides a vivid demonstration of the power of interactions. It normally involves just two factors:

- Liquid: tap water versus carbonated
- Solid: a peanut versus a raisin

Only one out of the four possible combinations produces an effect. Peanuts will generally float, and raisins usually sink in water. Peanuts are even more likely to float in carbonated liquid. However, when you drop in a raisin, the results are delightful. Raisins usually drop to the bottom. There they become coated with tiny bubbles, which lift the raisin back to the surface. The bubbles pop and the up-and-down process continues. Try it yourself at home. Show your family and friends and follow up by getting their ideas on the cause of this unexpected interaction of factors.

You must be somewhat wary about doing this experiment, because it occasionally fails. By dropping many raisins the chances improve for at least one raisin to "dance" (rise and fall repeatedly), but there's no guarantee of success.

A number of factors have been suggested as causes for failure, e.g., the freshness of the raisins and the specific brand of carbonated liquid. Some people even suggested that a popcorn kernel would work better than a raisin. These

Table 6-1. Factors for initial DOE on dancing objects

Factor	Name	Low Level (−)	High Level (+)
A	Material of container	Plastic	Glass
B	Size of container	Small	Large
C	Liquid	Club soda	Lemon lime
D	Temperature	Room	Ice cold
E	Cap on container	No	Yes
F	Type of object	Popcorn	Raisin
G	Age of object	Fresh	Stale

and other factors listed in Table 6-1 became the subject of a two-level factorial design.

Factor G, the aging of the objects, was accelerated by baking under a 100-watt light bulb for 15 minutes.

The full two-level factorial for seven factors requires 128 runs. This can be computed as follows: 2 * 2 * 2 * 2 * 2 * 2 * 2 = 128. However, with this many factors it becomes convenient to use the scientific notation: 2^7. We chose the 1/16th fraction (2^{-4} in scientific notation), which requires only 8 runs (= $2^7 * 2^{-4} = 2^{7-4}$ in scientific notation). This is a minimal design with Resolution III. At each set of conditions, we rated the dancing performance of 10 objects on a scale of 0 to 10. Table 6-2 shows the results. The actual order was randomized within one block ("Blk") of runs.

Table 6-2. Results from first dancing-raisin experiment

Std	Blk	A	B	C	D	E	F	G	Dancing Rating
1	1	−	−	−	+	+	+	−	1.5
2	1	+	−	−	−	−	+	+	2.0
3	1	−	+	−	−	+	−	+	1.0
4	1	+	+	−	+	−	−	−	4.0
5	1	−	−	+	+	−	−	+	1.5
6	1	+	−	+	−	+	−	−	1.0
7	1	−	+	+	−	−	+	−	5.0
8	1	+	+	+	+	+	+	+	1.0

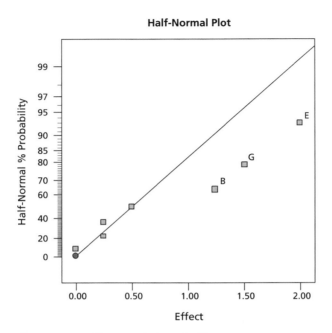

Figure 6-1. Half-normal plot of effects

The half-normal plot of effects is shown in Figure 6-1.

Three effects stood out: cap (E), age of object (G), and size of container (B). The ANOVA on the resulting model revealed highly significant statistics: an F-value of 27.78 with an associated probability of 0.0039, which falls far below the maximum threshhold of 0.05. (Reminder: a probability value of 0.05 indicates a 5% risk of a false positive, i.e., saying something happened that actually occurred due to chance. An outcome like this is commonly reported to be significant at the 95% confidence level.) The cube plot on Figure 6-2 shows the predicted responses for the three listed factors.

The worst rating (lowest) occurs at the upper, left, back corner of the cube. It reflects negative impacts of stale objects (G+) and capped liquid (E+), both of which make sense. However, the effect of size (B) does not make much sense. Could this be an alias for the real culprit, perhaps an interaction? Take a look at the alias structure for this Resolution III design, shown in Table 6-3. Each

GETTING THE MOST FROM MINIMAL-RUN DESIGNS

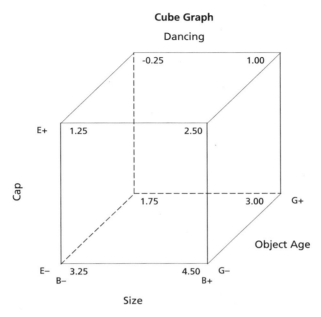

Figure 6-2. Cube plot of predicted responses

main effect is actually aliased with 15 other effects, but to simplify, we chose not to list interactions of three or more factors.

Can you pick out the likely suspect from the lineup for B? The possibilities are overwhelming, but they can be narrowed by assuming that the effects form a family.

Table 6-3. Alias structure for 2^{7-4} design (7 factors in 8 runs)

Labeled as	Actually
A	A + BD + CE + FG
B	B + AD + CF + EG
C	C + AE + BF + DG
D	D + AB + CG + EF
E	E + AC + BG + DF
F	F + AG + BC + DE
G	G + AF + BE + CD

> ### A VERY SCARY THOUGHT...
> Could a positive effect be cancelled by an 'anti-effect'?
>
> If you use a Resolution III design, be prepared for the possibility that a positive main effect may be wiped out by an aliased interaction of the same magnitude, but negative. The opposite could happen as well, or some combination of the above. Therefore, if nothing comes out significant from a Resolution III design, you cannot be certain that there are no active effects. Perhaps two or more big effects cancelled each other out.

The obvious alternative to B (size) is the interaction EG. However, this is only one of several alternative "hierarchical" models that maintain family unity:

- E, G, and EG (disguised as B)
- B, E, and BE (disguised as G)
- B, G, and BG (disguised as E)

The three interactions listed above are shown in Figure 6-3.

Notice that all of the interactions predict the same maximum outcome. However, the actual cause remains murky. The EG interaction remains far more plausible than the alternatives, but further experimentation is needed to clear things up.

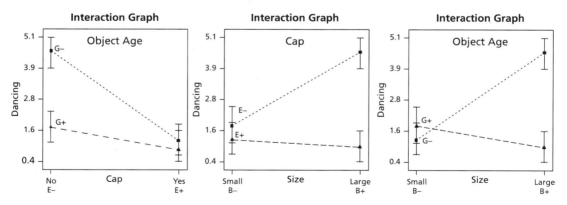

Figure 6-3 (a, b, c). Alternative interactions EG, BE, and BG (left to right)

Complete Fold-Over of Resolution III Design

By adding a second block of runs with signs reversed on all factors, you can break the aliases between main effects and two-factor interactions. This procedure is called a fold-over. It works on any Resolution III factorial. It's especially popular with Plackett-Burman designs, such as the 11 factors in 12-run choice. Let's see how the fold-over method works on the dancing raisin experiment. Table 6-4 shows the second block of experiments with all signs reversed on the control factors.

Notice that the signs of the two-factor interactions do not change from block 1 to block 2. For example, in block 1 the signs of columns B and EG are identical, but in block 2 they differ; thus, the combined design no longer aliases B with EG. If B really is the active effect, it should come out on the plot of effects for the combined design.

Table 6-4. Second block of runs after complete fold-over (selected interactions included)

Std	Blk	A	B	C	D	E	F	G	AD	BE	BG	EG	Dancing Rating
1	1	−	−	−	+	+	+	−	−	−	+	−	1.5
2	1	+	−	−	−	−	+	+	−	+	−	−	2.0
3	1	−	+	−	−	+	−	+	+	+	+	+	1.0
4	1	+	+	−	+	−	−	−	+	−	−	+	4.0
5	1	−	−	+	+	−	−	+	−	+	−	−	1.5
6	1	+	−	+	−	+	−	−	−	−	+	−	1.0
7	1	−	+	+	−	−	+	−	+	−	−	+	5.0
8	1	+	+	+	+	+	+	+	+	+	+	+	1.0
9	2	+	+	+	−	−	−	+	−	−	+	−	1.0
10	2	−	+	+	+	+	−	−	−	+	−	−	1.0
11	2	+	−	+	+	−	+	−	+	+	+	+	4.5
12	2	−	−	+	−	+	+	+	+	−	−	+	1.5
13	2	+	+	−	−	+	+	−	−	+	−	−	0.5
14	2	−	+	−	+	−	+	+	−	−	+	−	1.5
15	2	+	−	−	+	+	−	+	+	−	−	+	1.0
16	2	−	−	−	−	−	−	+	+	+	+	+	4.5

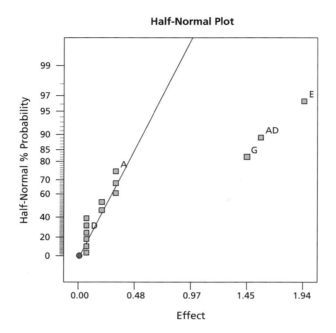

Figure 6-4. Half-normal plot of effects after fold-over

As you can see in Figure 6-4, factor B has disappeared and AD has taken its place. What happened to family unity?

The problem is that a complete fold-over of a Resolution III design does not break the aliasing of two-factor interactions, so AD remains aliased with EG as well as CF. The listing of the effect as AD—the interaction of container material with beverage temperature—is done arbitrarily by alphabetical order. Figure 6-5 shows the AD interaction. It makes no sense physically for the effect of material (A) to depend on temperature of the beverage (room temperature, D–, versus ice-cold, D+).

It's not so easy to discount the CF interaction: liquid type versus object type. However, the interaction between E and G is the most plausible, particularly since we now know that these two factors are present as main effects. Figure 6-6 shows the EG interaction.

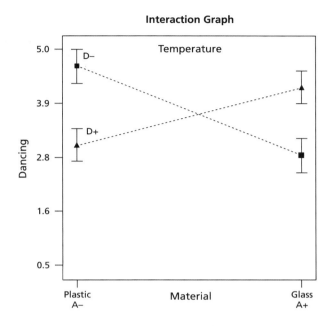

Figure 6-5. Interaction plot for AD

It appears that the effect of cap (E) depends on the age of the object (G). When the object is stale (the G+ line at the bottom of Figure 6-6), twisting on the bottle cap (going from E– at left to E+ at right) makes little difference. However, when the object is fresh (the G– line at the top), the bottle cap quenches the dancing reaction. More experiments are required to confirm this interaction. One obvious way to do this is to conduct a full factorial on E and G alone. Other ideas on de-aliasing are presented in the next sidebar.

> ### AN ALIAS BY ANY OTHER NAME IS NOT NECESSARILY THE SAME
>
> You might be surprised that aliased interactions such as AD and EG do not look alike. Their coefficients are identical, but the plots differ because they combine the interaction with their parent terms. If you still don't believe it, construct the interaction plot for CF using the methods described in Chapter 3. This interaction is also aliased with AD. By comparing the graphs of aliased interactions (such as AD versus EG versus CF,) you can make an educated guess as to which make the most sense to investigate further.

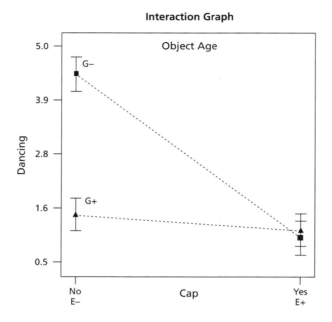

Figure 6-6. Plot of interaction EG

Single Factor Fold-Over

Another simple way to de-alias a Resolution III design is the "single-factor fold-over." Like a complete fold-over, you must do a second block of runs, but in this variation of the general method, you change signs on only one factor. This factor and all its two-factor interactions become clear of any other main effects or interactions. For example, let's go back to the original Resolution III design on the dancing raisins. It makes sense to focus on the biggest effect: E (refer to Figure 6-2). The end result of the fold-over on factor E (only) is a design with 16 runs in two blocks of eight. The resulting alias structure (main effects and two-factor interactions only) is shown in Table 6-5.

The combined design remains a Resolution III because, with the exception of E, all main effects remained aliased with two 2-factor interactions. Factor E is a Resolution V, because the main effect is clear of aliases, and the two-factor interactions are aliased only with three-factor interactions. In the case of the dancing raisins, the single-factor fold-over would have revealed that B actu-

Table 6-5. Alias structure for 2^{7-4} design after fold-over on factor E

Labeled as	Actually
A	A + BD + FG
B	B + AD + CF
C	C + BF + DG
D	D + AB + CG
E	E
F	F + AG + BC
G	G + AF + CD
AC	AC + BG + DF
AE, BE, CE, DE, EF, EG	AE, BE, CE, DE, EF, EG

ally was EG, not AD. On the other hand, factor G remains aliased with two 2-factor interactions. You can't win either way.

> **FOLD-OVER DOESN'T ALWAYS ADD A NEW WRINKLE**
>
> The complete fold-over of Resolution IV designs may do nothing more than replicate the design so that it remains Resolution IV. This would happen if you folded the 16 runs in Table 6-4. By folding only certain columns of a Resolution IV design, you might succeed in de-aliasing some of the two-factor interactions. Other than trying different combinations of columns to fold-over, the only sure way to eliminate aliases is the single-factor fold-over, which works on Resolution IV the same as Resolution III designs: the main effect and all the two-factor interactions of the factor you choose will be cleared.

Choose a High-Resolution Design to Reduce Aliasing Problems

The best solution remains to run a higher resolution design by selecting fewer factors and/or a bigger design. For example, you could run seven factors in 32 runs (2^{7-2}). This option was discussed in the previous chapter on fractional factorials. It is a Resolution IV design, but all 7 main effects and 15 of the 21 two-factor interactions are clear of other two-factor interactions. The remaining 6 two-factor interactions are shown in Table 6-6.

Table 6-6. Problem aliases for 2^{7-2} design (7 factors in 32 runs)

Labeled as	Actually
DE	DE + FG
DF	DF + EG
DG	DG + EF

The trick is to label the likely interactors anything but D, E, F, and G. For example, knowing now that capping and age interact in the dancing raisin experiment, we would not label these factors E and G. If only we knew then what we know now!

Practice Problems

PROBLEM 6-1

(Warning: the following scenario contains liberal doses of fantasy.) One of the authors became envious of the skating ability of his co-author. The 'wannabe' skater secretly got together with a local manufacturer of in-line skates and borrowed the latest and greatest experimental gear. Bewildered by all the options, he decided to try various combinations at the local domed stadium, which opened its concourse to skaters when not in use for sporting events. The special skates would have to be returned fairly quickly, so a quick-and-dirty fractional factorial was set up. If anything proved to be statistically significant, the result would be a faster time around the track (with ego-gratification to the skater). The experimenter knew that due to aliasing of effects, he would not get a true picture of what really enhanced speed. This could be calculated later via followup designs using the fold-over method. Table 6-7 shows the initial design, a 2^{7-4} Resolution III fractional factorial, and the resulting times around the track.

Here's more background on the factors and levels to help you interpret the outcome:

    A. Pad goes inside skate to elevate the heel: Out −, In +

    B. Bearing constructed either from old material (−) or new high-tech alloy (+)

    C. Gloves made specially for in-line skating to protect wrists: On −, Off +

Table 6-7. First experiment on in-line skates

Std	A: Pad	B: Bearing	C: Gloves	D: Helmet	E: Wheels	F: Covers	G: Neon	Time (sec.)
1	Out	Old	On	Front	Soft	Off	Off	195
2	In	Old	On	Back	Hard	Off	On	192
3	Out	New	On	Back	Soft	On	On	200
4	In	New	On	Front	Hard	On	Off	165
5	Out	Old	Off	Front	Hard	On	On	190
6	In	Old	Off	Back	Soft	On	Off	195
7	Out	New	Off	Back	Hard	Off	Off	166
8	In	New	Off	Front	Soft	Off	On	201

D. Helmet fits with logo to back (–) or front (+)—can't tell which is correct, so try both and ignore laughs when wrong!

E. Wheels can be made of hard (–) or soft (+) polymer

F. Covers go on wheels to make them look faster: On –, Off +

G. Neon lighting (from generator on skates) for night-time use: Off –, On +

Analyze the data to see if any of these factors appear to be significant. Do the results make sense? Could the real answer be disguised by an alias (hint: refer to Table 6-3)? (Suggestion: use the software provided with this book. Create a two-level factorial design for seven factors in 8 runs and sort the resulting layout by standard order. Then enter the time data from Table 6-7. Do the analysis as outlined in the factorial tutorial that comes with the program.)

PROBLEM 6-2

This is a continuation of the skating saga from Problem 6-1. Rolling right along, the experimenter decides to do a complete fold-over on the initial design. Table 6-8 shows the factor levels and results for this followup design.

Add these data to the design from Problem 6-1, and analyze it as a second block. Do any of the significant model terms turn out to be interactions, rather than main effects? Remember that the fold-over upgrades the Resolution III design to Resolution IV, but interactions may still be aliased with

Table 6-8. Followup design (fold-over) to initial DOE on in-line skates

Std	A: Pad	B: Bearing	C: Gloves	D: Helmet	E: Wheels	F: Covers	G: Neon	Time (sec.)
9	In	New	Off	Back	Hard	On	On	175
10	Out	New	Off	Front	Soft	On	Off	211
11	In	Old	Off	Front	Hard	Off	Off	202
12	Out	Old	Off	Back	Soft	Off	On	205
13	In	New	On	Back	Soft	Off	Off	212
14	Out	New	On	Front	Hard	Off	On	175
15	In	Old	On	Front	Soft	On	On	204
16	Out	Old	On	Back	Hard	On	Off	201

other interactions. If interactions do appear, do they make sense versus the aliased alternatives? (Suggestion: use the software provided with this book. Look for a data file called "Prb6-2-skate2," open it, then do the analysis. View the aliased interactions and try substitutions. Compare the alternatives graphically, in a manner similar to that outlined for the dancing-raisin case.)

CHAPTER 7
GENERAL FACTORIAL DESIGNS

> *"Invention is discernment, choice...*
> *The useful combinations are precisely the most beautiful."*
> —HENRI POINCARÉ

In Chapter 2, we showed you how to do simple comparative experiments on individual factors with any number of levels. We then shifted to experiments with multiple factors, but restricted to just two levels. You might find it helpful at this point to refer to the flowchart in the Introduction. It directs you to the various types of experiments covered in this book, depending on:

- Number of factors,
- Number of levels, and
- Nature of factors—numerical or categorical, process, or mixture.

In many situations, you find yourself confronted with a number of categorical alternatives, such as three suppliers (A versus B versus C), plus other factors that could interact. For example, you might be interested in brewing several varieties of coffee with different flavoring additives. A simple solution to this problem is to run all combinations of all factors. We call this a "general factorial" design. These designs, also called "mixed factorials," are not very popular because:

1. The number of combinations becomes excessive after only a few factors. It may be possible to perform fractional designs, but except for special cases documented in the literature, you must resort to computer-aided selection via matrix-based methods. This goes well beyond the scope of *DOE Simplified*.

2. Each design is unique to the given situation, with many possible levels for the specified varying number of factors. Therefore, you may not find a template or example that you can follow for design and analysis.

3. Due to the already large number of combinations, general factorials for three or more factors are often done without replication. Because no pure error is available, you must assume that highest-order interactions are insignificant and go from there to do the ANOVA.
4. You can't use the normal plots for effect selection.
5. Predictive models for categorical factors must be coded in a non-intuitive manner.

To avoid these complications, stick with simple comparisons and two-level factorials. For example, you could first screen the several varieties of coffee via a one-factor, simple-comparative taste test. Then after eliminating all but the top two varieties, combine these with your two favorite flavors, plus other factors at two levels, in a standard factorial. However, this simplistic approach to experiment design may overlook critical interactions that will be revealed in a general factorial design. This more comprehensive choice will be illustrated in the following case study.

Putting a Spring in Your Step—A General Factorial Design on Spring Toys

A coiled spring, made to specific dimensions (see sidebar), will gracefully "walk" down an incline. The most obvious factor affecting speed is the degree of incline, which must be between 20 and 40 degrees. If the incline is too shallow, the coil will not move. Too steep an incline causes the coil to tumble or roll out of control. Friction also plays a role. During our experiment, after observing slippage on bare wood, we added a high-friction rubber mat to our surface. Many other variables are associated with the construction of the coil, such as the spring constant, mass, diameter, and height.

Several varieties of coiled springs are made by James Industries, including the traditional metal Slinky and the Slinky Junior—a smaller version made out of styrene-butadiene plastic. These are the first two bent springs shown in Figure 7-1 from bottom to top. We also looked at a larger plastic Slinky (shown in the hands of tester in Figure 7-1), but it walked too slowly and stopped frequently.

GENERAL FACTORIAL DESIGNS

> ## THE ONE PERSON WHO KNEW IT WAS A SLINKY
>
> Richard James, a naval engineer, invented the Slinky. During World War II he worked on springs for keeping sensitive instruments steady at sea. One day he accidentally knocked an experimental spring off a table on to a pile of books. It tumbled each step of the way in a delightful walking motion. After seeing the reaction of neighborhood kids to his new toy, James decided to develop it commercially. The first Slinky hit the Philadelphia market in 1945. It became an instant success. The Slinky, now made by James Industries of Hollidaysburg, Pennsylvania, remains popular not only with children, but also with physics teachers who use the toy to illustrate wave properties and energy states. Here's a bit of trivia you can try on your friends: the original Slinky measures 87 feet when stretched.
>
> *"Isn't this incredible! It's gotta be some kind of record!"*
> —Ace Ventura (*When Nature Calls*) after his Slinky
>   walks down a mountainous stairway

Figure 7-1. Various spring toys on high-friction rubber mat

Table 7-1. General factorial design (replicated) on spring toys

Std	A: Spring Toy	B: Incline	Time (seconds)
1, 2	Metal Slinky	Shallow	5.57, 5.75
3, 4	Slinky Junior	Shallow	5.08, 5.36
5, 6	Generic plastic	Shallow	3.03, 3.34
7, 8	Metal Slinky	Steep	4.67, 4.95
9, 10	Slinky Junior	Steep	4.23, 4.98
11, 12	Generic plastic	Steep	3.58, 4.50

Two additional, generic, plastic spring toys (the coiled ones in the picture) were tested. The smaller of these walked too fast for valid comparison to the Slinky brands. That left three coiled spring toys to test at two inclines in a full factorial experiment. We replicated each of the six combinations (3×2) in a completely randomized test plan. The 12 results are shown sorted by standard order in Table 7-1.

The response is time in seconds for the springs to walk a four-foot inclined plank. The two results per cell represent the replicated runs. As noted earlier, these were done at random, not one after the other, as shown. Therefore, the difference in time reflects variations due to setting of the board, placement of the coil, how the operator made it move, and so forth. The analysis of variance (ANOVA) is seen in Table 7-2. Procedures for calculating the sums of squares are provided in referenced textbooks and encoded in many statistical software packages.

Table 7-2. ANOVA for general factorial on spring toys

Source	Sum of Squares	Df	Mean Square	F Value	Prob >F
Model	7.73	5	1.55	10.96	0.0056
A	5.90	2	2.95	20.90	0.0020
B	0.12	1	0.12	0.88	0.3848
AB	1.71	2	0.85	6.05	0.0365
Residual	0.85	6	0.14		
Cor Total	8.58	11			

GENERAL FACTORIAL DESIGNS

Look first at the probability value (Prob >F) for the model. Recall that we consider values less than 0.05 to be significant. (Reminder: a probability value of 0.05 indicates a 5% risk of a false positive, i.e., saying something happened that actually occurred due to chance. An outcome like this is commonly reported to be significant at the 95% confidence level.) In this case, the probability value of 0.0056 for the model easily passes the test for significance, falling well below the 0.05 threshold. Because we replicated all six combinations, we obtained six degrees of freedom (df) for estimation of error. This is shown in the line labeled "Residual." The three effects can be individually tested against the residual. The result is an unexpectedly large interaction (AB). This is shown graphically in Figure 7-2.

The two Slinky springs behaved as expected by walking significantly faster down the steeper incline (B+). However, just the opposite occurred with the generic plastic spring. It was observed taking unusually long but low steps down the shallow slope. It didn't look very elegant, but it was fast. On the steep slope the generic spring did the more pleasing, but slower and shorter, high step, like the Slinky toys.

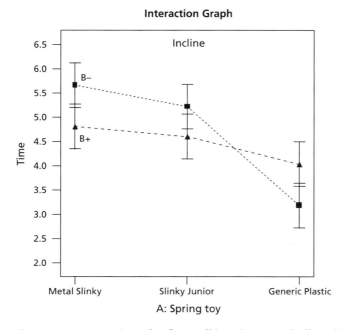

Figure 7-2. Interaction plot for walking times at shallow (B–) versus steep (B+) incline

How to Analyze Unreplicated General Factorials

The example shown above involved only six combinations for the full factorial, so doing a complete replicate of the design was very feasible. When you add a third factor, the cost of replication goes up by two-fold or more, depending on how many levels you must run. In this case, it's common practice to assume that interactions of three or more factors are not significant. In other words, whatever effect you see for these higher order interactions are more than likely due to normal process variability. Therefore, these effects can be used for estimating error. Other insignificant interactions may also be pooled as error in much the same way as with two-level factorials earlier in this book. Let's look at an example.

Before conducting the earlier case study on the springs, we rejected two abnormal toys:

- A larger one—the "original plastic Slinky."
- A smaller one—a generic plastic variety.

We will go back and test these two springs, plus two from the earlier test, excluding the generic plastic toy. The factor of incline will be retained at two levels. A third factor will be added: whether an adult or a child makes the springs walk. The resulting 16 combinations (4×2×2) are shown in Table 7-3, along with the resulting times.

The impact of the three main effects and their interactions can be assessed via analysis of variance. In general factorials like this, it helps to view a preliminary breakdown of sum of squares (a measure of variance) before doing the actual ANOVA. In many cases you will see that some or all of the two-factor interactions contribute very little to the overall variance. These trivial effects can then be used as estimates of error, in addition to the three-factor or higher interactions already destined for error. Table 7-4 provides the sum of squares, degrees of freedom, and the associated mean squares for the time data from the experiment on spring toys.

The main effect from factor A ranks first on the basis of sum of squares. It accounts for 73.1% (the "contribution") of the total sum of squares. The main effect of B comes next, at 11.4% contribution. The other main effect, from factor C, contributes relatively little to the variance: only 1.3%. Continuing down the list, you see fairly low contributions from each of the three two-

Table 7-3. Second experiment on spring toys

Std	A: Spring Toy	B: Incline	C: Operator	Time (seconds)
1	Metal Slinky	Shallow	Child	5.57
2	Slinky Junior	Shallow	Child	5.08
3	Plastic Slinky	Shallow	Child	6.37
4	Small generic	Shallow	Child	3.03
5	Metal Slinky	Steep	Child	4.67
6	Slinky Junior	Steep	Child	4.23
7	Plastic Slinky	Steep	Child	4.70
8	Small generic	Steep	Child	3.28
9	Metal Slinky	Shallow	Adult	6.51
10	Slinky Junior	Shallow	Adult	5.21
11	Plastic Slinky	Shallow	Adult	6.25
12	Small generic	Shallow	Adult	3.47
13	Metal Slinky	Steep	Adult	4.88
14	Slinky Junior	Steep	Adult	3.39
15	Plastic Slinky	Steep	Adult	6.72
16	Small generic	Steep	Adult	2.80

Table 7-4. Breakdown of variance for second experiment on spring toys

Term	Sum of Squares	Contribution	DF	Mean Square
A	18.68	73.1%	3	6.23
B	2.91	11.4%	1	2.91
C	0.33	1.3%	1	0.33
AB	0.88	3.4%	3	0.29
AC	1.03	4.0%	3	0.34
BC	0.014	0.1%	1	0.014
ABC	1.71	6.7%	3	0.57
Total	25.55	100.0%	15	

factor interactions (AB, AC, BC). Finally you get to the three-factor interaction effect ABC, which, as stated earlier, will be used as an estimate of error. Since none of the other interactions caused any larger sum of squares, these will also be thrown into the error pool (labeled "Residual" in the ANOVA). The rankings in terms of mean square (the last column in Table 7-4) provide support for this decision. Although it appears to be trivial, the C term will be kept in the model because it's a main effect. The resulting main-effects-only (A, B, C) model is the subject of the ANOVA shown in Table 7-5.

The model passes the significance test with a probability of less than 0.001 (>99.9% confidence). As expected, A and B are significant, as indicated by their probability values being less than 0.05. By this same criteria, we conclude that the effect of factor C (operator: child versus adult) is not significant. The probability is high (0.3626) that it occurred due to chance. In other words, it makes no difference whether the adult or the child walked the spring toy. The effect of A—the type of spring toy—can be seen in Figure 7-3, the interaction plot for AB. (Normally we wouldn't show an interaction plot when the model contains only main effects, but it will be helpful in this case for purposes of comparison.)

Compare Figure 7-3 with Figure 7-2. Notice that in this second design the lines are parallel, which indicates the absence of an interaction. The effects of A and B do not depend on each other. You can see from Figure 7-3 that the shallower incline (B-) consistently produced longer times. The main-effect plot in Figure 7-4 also shows the effect of B, but without the superfluous detail on the type of spring.

Table 7-5. ANOVA for selected model (main effects only)

Source	Sum of Squares	Df	Mean Square	F Value	Prob >F
Model	21.92	5	4.38	12.06	0.0006
A	18.68	3	6.23	17.14	0.0003
B	2.91	1	2.91	8.00	0.0179
C	0.33	1	0.33	0.91	0.3626
Residual	3.63	10	0.36		
Cor Total	25.55	15			

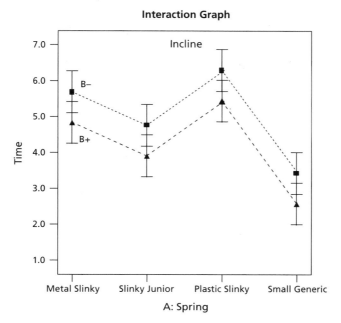

Figure 7-3. Plot for walking times at shallow (B–) versus steep (B+) incline

> ### *QUANTIFYING THE BEHAVIOR OF SPRING TOYS*
> A more scientific approach to understanding the walking behavior of the spring toys is to quantify the variables. To do this, you would need a spring-making machine that would allow you to make coils, of varying height and diameter, from plastics with varying physical properties. The general factorial indicates that walking speeds increase as the spring diameter decreases, but the exact relationship remains obscure. Similarly, the impact of incline is known only in a general way. By quantifying the degree of incline, a mathematical model can be constructed. We provide an overview of ways to do this in Chapter 8, on "response surface methods" (RSM).

As expected, the spring toys walked faster down the steeper slope.

This example created a template for general factorial design. Setup is straightforward: lay out all combinations by crossing the factors. With the aid of computers, you can create a fractional design that fits only your desired model terms, but the procedure to do so is beyond the scope of this book.

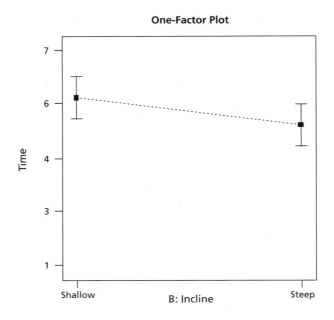

Figure 7-4. Main effect plot for incline

Check your software to see if it supports this feature, called "optimal design." However, if you don't include too many factors or go overboard on the number of levels, go ahead and do the full factorial. Replication is advised for the two-factor design, but it may not be realistic for larger factorials. In that case, you must designate higher-order interactions as error and hope for the best.

If at all possible, try to redesign your experiment to fit a standard two-level factorial. This is a much simpler approach. Another tactic is to quantify the factors and do a response surface design (see related sidebar).

Practice Problems

PROBLEM 7-1

Montgomery describes a general factorial design on battery life (see referenced textbook *Design and Analysis of Experiments*, p. 235). Three materials are evaluated at three levels of temperature. Each experimental combination is replicated

GENERAL FACTORIAL DESIGNS

Table 7-6. General factorial on battery
(response is life in hours)

Material Type	Temperature (Deg F)					
	15		70		125	
A1	130	155	34	40	20	70
	74	180	80	75	82	58
A2	150	188	136	122	25	70
	159	126	106	115	58	45
A3	138	110	174	120	96	104
	168	160	150	139	82	60

four times in a completely randomized design. The responses from the resulting 36 runs can be seen in Table 7-6.

Which, if any, material significantly improves life? How are results affected by temperature? Make a recommendation for constructing the battery. (Suggestion: use the software provided with this book. First do the tutorial on general factorials that comes with the program. It's keyed to the data in Table 7-6.)

PROBLEM 7-2

One of the authors lives about 20 miles due east of his workplace in Minneapolis. He can choose among three routes: southern loop via freeway; directly into town via stop-lighted highway; or northern loop on freeway. His work hours are flexible, so he can come in either early or late. (His partners frown on him coming in late and leaving early, but they don't object if he comes in early and leaves late!) Aside from these two factors, the time of his commute changes from day to day depending on weather conditions and traffic problems. Therefore, he decided that the only sure method for determining the best route and time would be to conduct a full factorial replicated at least twice. Table 7-7 shows the results from this general factorial design. The actual run order was chosen at random.

Table 7-7. Commute times for different routes at varying schedules

Std	A: Route	B: Depart	Commute (minutes)
1, 2	South	Early	27.42, 29.1
3, 4	Central	Early	29.02, 28.71
5, 6	North	Early	28.51, 27.43
7, 8	South	Late	33.57, 32.91
9, 10	Central	Late	30.71, 29.12
11, 12	North	Late	29.8, 30.94

Analyze this data. Does it matter which route the driver takes? Is it better to leave early or late? Does the route depend on the time of departure? (Suggestion: use the software provided with this book. Look for a data file called "Prb7-2-drive," open it, then do the analysis. View the interaction graph. It will help you answer the questions above.)

CHAPTER 8
RESPONSE SURFACE METHODS FOR OPTIMIZATION

"The first step comes in a burst. Then 'bugs,' as such little faults and difficulties are called, show themselves. Months of intense study and labor are required before commercial success."

—Thomas Edison (1878)

In this chapter, we will provide a broad overview of more advanced techniques for optimization called "response surface methods" (RSM). RSM should be applied only after completing the initial phases of experimentation:

1. Fractional two-level designs, which screen the vital few from the trivial many factors.
2. Full factorials, which study the vital few factors in depth and define the region of interest.

The goal of RSM is to generate a map of response, either in the form of contours or as a 3-D rendering. These maps are much like those used by a geologist for topography, but instead of elevation they show your response, such as process yield. Figure 8-1 shows examples of response surfaces. The surface on the left exhibits a "simple maximum," a very desirable outcome because it reveals the peak of response. The surface on the right, called a saddle point, is much more complex. It exhibits two maximums. You may encounter other types of surfaces, such as simple minimums or rising ridges.

A two-level design cannot fit the surfaces shown in Figure 8-1, but it can detect the presence of curvature with the addition of "centerpoints." We will show how to add centerpoints, and, if curvature is significant, how to augment your design into an RSM.

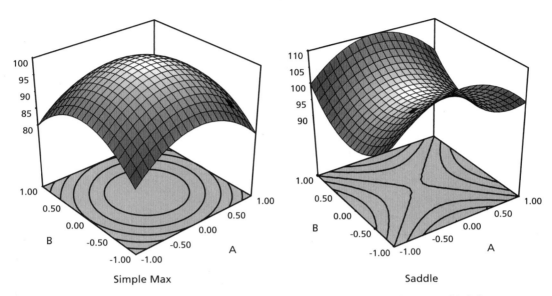

Figure 8-1 (a, b). Response surfaces—simple maximum (left) and saddle point (right)

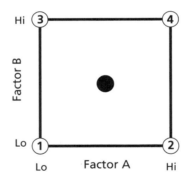

Figure 8-2. Two-level factorial design with centerpoint(s)

Centerpoints Detect Curvature in Confetti

Centerpoints are created by setting all factors at their midpoints. In coded form, centerpoints fall at the all-zero level. For example, look at an experiment on confetti—the strips of paper that people throw into the air to celebrate a big event.

The objective is to cut strips that drop slowly through the air. If the dimensions are optimal, the confetti spins and moves at random angles that please the eye. Table 8-1 shows the factors and levels to be tested. Note the addition

RESPONSE SURFACE METHODS FOR OPTIMIZATION

Table 8-1. Two-level factorial with centerpoints for confetti

Factor	Name	Units	Low Level (−)	Center (0)	High Level (+)
A	Width	Inches	1	2	3
B	Height	Inches	3	4	5

of centerpoints (coded as 0). This is a safety measure that plugs the gap between low (−) and high (+) levels.

For convenience of construction, the confetti specified above is larger and wider than that available commercially. The actual design is shown in Table 8-2. We replicated the centerpoint four times to provide more power for the analysis. These points, along with all the others, were performed in random order. The centerpoints act as a barometer of the variability in the system.

> ### TRUE REPLICATION VERSUS REPEAT MEASUREMENT
>
> It often helps to repeat response measurements many times for each run. For example, in the confetti experiment, each strand was dropped 10 times and then averaged. However, this cannot be considered a true replicate because some operations were not repeated, such as the cutting of the paper. In the same vein, when replicating centerpoints, you must repeat all the steps. For example, it would have been easy just to reuse the 2 by 4 inch confetti, but we actually re-cut to this dimension four times. Therefore, we obtained an accurate estimate of the "pure error" of the confetti production process.

Table 8-2. Design layout and results for confetti experiment

Std	A: Width (inches)	B: Length (inches)	Time (seconds)
1	1.00	3.00	2.5
2	3.00	3.00	1.9
3	1.00	5.00	2.8
4	3.00	5.00	2.0
5	2.00	4.00	2.8
6	2.00	4.00	2.7
7	2.00	4.00	2.6
8	2.00	4.00	2.7

> ### ADDING A CENTERPOINT DOES NOT CREATE A FULL THREE-LEVEL DESIGN
>
> The two-level design with centerpoint(s) (pictured in Figure 8-2) requires all factors to be set at their midlevels, around which you run only the combinations of the extreme lows and highs. It differs from a full three-level factorial, which requires nine distinct combinations, including points at the midpoints of the edges. The two-level factorial with centerpoint(s) will reveal curvature in your system, but it does not provide the complete picture that would be obtained by doing the full three-level factorial.

The response is flight time in seconds from a height of five feet. The half-normal plot of effects for this data is shown in Figure 8-3.

Factor A, the width, stands out as a very large effect. On the other end of the effect scale (nearest zero), notice the three triangular symbols. These come from the replicated centerpoints, which contribute three degrees of freedom for estimation of "pure error." In line with the pure error, you will find the

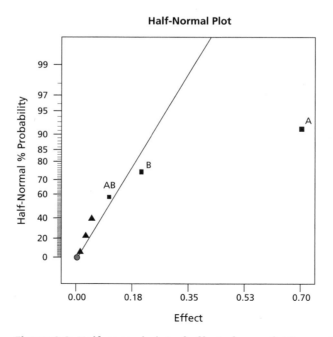

Figure 8-3. Half-normal plot of effects for confetti experiment

Table 8-3. ANOVA for confetti experiment (effects B and AB used for lack of fit test)

Source	Sum of Squares	Df	Mean Square	F Value	Prob >F
Model	0.49	1	0.49	35.00	0.0020
A	0.49	1	0.49	35.00	0.0020
Curvature	0.32	1	0.32	22.86	0.0050
Residual	0.070	5	0.014		
Lack of Fit	0.050	2	0.025	3.75	0.1527
Pure Error	0.020	3	0.0067		
Cor Total	0.88	7			

main effect of B (length) and the interaction AB. These two relatively trivial effects will be thrown into the residual pool under a new label, "lack of fit," to differentiate these estimates of error from the "pure error." (Details on lack of fit can be found in the sidebar on this topic.) The pure error is included in the residual subtotal in the ANOVA, shown in Table 8-3, which also exhibits a new row, "Curvature."

Apply the usual 0.05 rule to assess the significance of the curvature. In this case, the probability value of 0.005 for curvature falls below the acceptable threshold of 0.05, so it cannot be ignored. That's bad. It means that the results at the centerpoint were unexpectedly high or low relative to the factorial points around it. Figure 8-4 shows effect plots of the response versus factors A and B.

The relationships obviously are not linear. Notice that the centerpoint responses stay the same in both plots. Because all factors are run at their midlevels, we cannot say whether the observed curvature occurs in the A or the B direction, or some of both. Statisticians express this confusion as an alias relationship: curvature = $A^2 + B^2$. It will take more experimentation to pin this down. The next step is to augment the existing design via response surface methods (RSM).

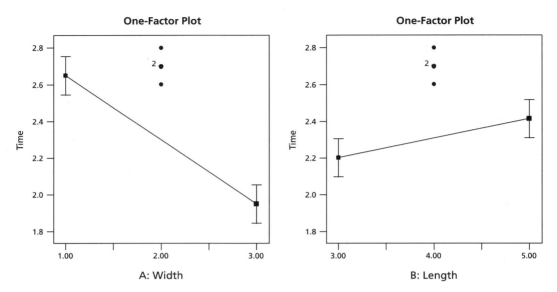

Figure 8-4 (a, b). Effect plots of time of confetti flight versus factors A and B

LACK OF FIT MAY BE FASHIONABLE, BUT IT IS NOT DESIRABLE FOR EXPERIMENTERS

You may have noticed a new line in the ANOVA called "lack of fit." This tests whether the model adequately describes the actual response surface. It becomes possible only when you include replicates in your design. The lack of fit test compares the error from excess design points (beyond what's needed for the model) versus the pure error from the replicates. As a rule of thumb, a probability value of 0.05 or less for the F-value indicates a significant lack of fit—which you don't want.

Augmenting to a Central Composite Design (CCD)

The remedy for dealing with significant curvature in two-level factorial design is to add more points. By locating the new points along the axes of the factor space, you can create a "central composite design"(CCD). If constructed properly, the CCD provides a solid foundation for generating a response surface map. Figure 8-5 shows the two-factor and three-factor CCDs.

For maximum efficiency, the "axial" (or "star") points should be located a specific distance *outside* the original factor range. The ideal location can be found

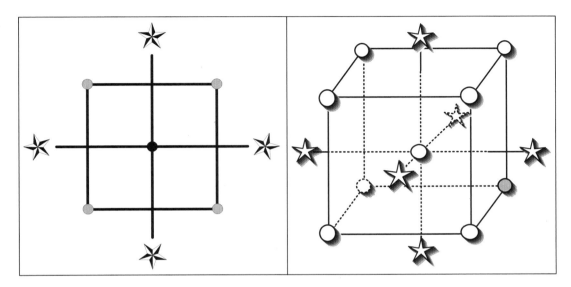

Figure 8-5 (a, b). Central composite designs for two and three factors, respectively

through research or calculated by software, but it will be very close to the square root of the number of factors. For example, for the two-factor design used to characterize confetti, the best place to add points is 1.4 coded units from the center. The augmented design is shown in Table 8-4. The new points are designated as block 2. Notice the additional centerpoints. These provide a link between the blocks and add more power to the estimation of second-order effects needed to characterize curvature.

The CCD contains five levels of each factor: low axial, low factorial, center, high factorial, and high axial. With this many levels, it generates enough information to fit a second-order polynomial called a "quadratic." Standard statistical software can compute the actual fitting of the model. The quadratic model for confetti flight time is:

$$\text{Time} = 2.68 - 0.30A + 0.12B - 0.050AB - 0.31A^2 + 0.020B^2$$

This model is expressed in terms of the coded factor levels shown in Table 8-1. The coding eliminates problems caused by varying units of measure, such as inches versus centimeters, which can create problems when comparing coefficients. In this case, the A-squared (A^2) term has the largest coefficient, which indicates curvature along this dimension. The ANOVA, shown in Table 8-5,

Table 8-4. Central composite design for confetti

Std	Block	Type	A: Width (inches)	B: Length (inches)	Time (seconds)
1	1	Factorial	1.00	3.00	2.5
2	1	Factorial	3.00	3.00	1.9
3	1	Factorial	1.00	5.00	2.8
4	1	Factorial	3.00	5.00	2.0
5	1	Center	2.00	4.00	2.8
6	1	Center	2.00	4.00	2.7
7	1	Center	2.00	4.00	2.6
8	1	Center	2.00	4.00	2.7
9	2	Axial	0.60	4.00	2.5
10	2	Axial	3.40	4.00	1.8
11	2	Axial	2.00	2.60	2.6
12	2	Axial	2.00	5.40	3.0
13	2	Center	2.00	4.00	2.5
14	2	Center	2.00	4.00	2.6
15	2	Center	2.00	4.00	2.6
16	2	Center	2.00	4.00	2.9

Table 8-5. ANOVA for CCD on confetti

Source	Sum of Squares	Df	Mean Square	F Value	Prob >F
Block	0.016	1	0.016		
Model	1.59	5	0.32	15.50	0.0003
A	0.71	1	0.71	34.72	0.0002
B	0.12	1	0.12	5.67	0.0412
A^2	0.75	1	0.75	36.49	0.0002
B^2	0.003	1	0.003	0.14	0.7201
AB	0.010	1	0.010	0.49	0.5032
Residual	0.19	9	0.021		
Lack of Fit	0.075	3	0.025	1.36	0.3402
Pure Error	0.11	6	0.018		
Cor Total	1.79	15			

RESPONSE SURFACE METHODS FOR OPTIMIZATION

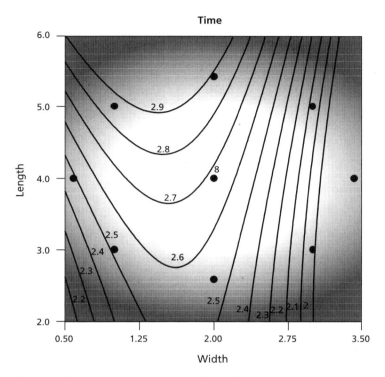

Figure 8-6. Contour graph for confetti flight time

indicates a high degree of significance for this term and the model as a whole. The AB and B^2 terms are not significant, but there is no benefit to eliminating them from the model because the response surface will not be affected.

Lack of fit is not significant (because the probability value of 0.3578 exceeds the threshold value of 0.05). Also, diagnosis of residuals showed no abnormality. Therefore, the model is statistically solid. The resulting contour graph is shown in Figure 8-6.

Each contour represents a combination of input factors that produces a constant response, as shown by the respective labels. The actual runs are shown as dots. (The number 8 by the centerpoint indicates the number of replicates at this set of conditions. In other words, at eight random intervals throughout the experiment, we reproduced confetti with the midpoint dimensions of 2 by 4 inches.) Normally we would restrict the axes to the factorial range to avoid extrapolation beyond the experimental space, but we wanted to show

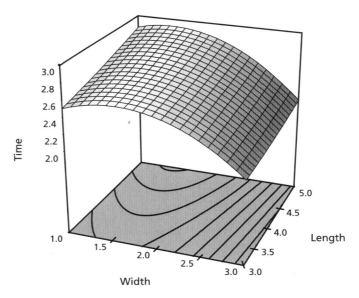

Figure 8-7. Response surface for confetti flight time

the entire design. Notice the darkening in areas outside of the actual design space, especially in the corners. These represent areas where predictions will be unreliable due to lack of information. Figure 8-7 shows a 3-D response surface with the ranges reduced to their proper levels.

The maximum flight time within this factorial range occurs at a width of 1.44 inches and length of 5 inches. Longer confetti might fly even longer, but this would need to be determined via further experimentation.

> ### WHERE'S THERE'S SMOKE, THE PROBABILITY IS HIGH THERE'S A FIRE
>
> A scientist, engineer, and statistician watched their research lab burn down as a result of combustion of confetti in a paper shredder. They speculated as to the cause. "It's an exothermic reaction," said the scientist. "That's obvious," replied the engineer, "What's really important is the lack of heat transfer due to inadequate ventilation." The two technical professionals then turned to their statistical colleague, who said "I have no idea what caused the fire, but I can advise that you do some replication: Let's burn down another lab and see what happens!"

CHAPTER 9
MIXTURE DESIGN

"The good things of life are not to be had singly, but come to us with a mixture."

—CHARLES LAMB

The cliché image of experimentation is a crazed person in a lab coat pouring fluorescent liquids into bubbling beakers. Ironically, the standard approaches for DOE don't work very well on experiments that involve mixtures. For example, what would happen if you performed a factorial design on combinations of lemonade and apple juice? Table 9-1 shows the experimental layout with two levels of each factor—either one or two cups.

Notice that standard orders 1 and 4 call for mixtures with the same ratio of lemonade to apple juice. The total amount varies, but will have no effect on responses such as taste, color, or viscosity. Therefore, it makes no sense to do the complete design. When responses depend only on proportions and not the amount of ingredients, factorial designs don't work very well.

Table 9-1. Factorial design on mixture of fruit juices

Std Order	A: Lemonade (cups)	B: Apple Juice (cups)	Ratio
1	1	1	1/1
2	2	1	2/1
3	1	2	1/2
4	2	2	1/1

> **SHOTGUN APPROACH TO MIXTURE DESIGN**
>
> Chemists are famous for creating mysterious concoctions seemingly by magic. A typical example is Hoppe's Nitro Powder Solvent Number 9, invented by Frank August Hoppe. While fighting in the Spanish-American War, Captain Hoppe found it extremely difficult to ream corrosion from his gun barrel. The problem was aggravated by antagonistic effects from mixing old black powder with new smokeless powder. After several years of experimenting in his shed, Hoppe came up with a mixture of nine chemicals that worked very effectively. A century or so later, his cleaning solution is still sold. The composition remains a trade secret.

Another approach to this problem is to take the variable of amount out of the experiment and work on a percentage basis. Table 9-2 shows the layout for a second attempt at the juice experiment, with each "component" at two levels: 0 or 100 percent.

This design obviously doesn't work either, because it asks for impossible totals. It illustrates a second aspect of mixtures: the total is "constrained" in that the ingredients must add up to 100 percent.

Two-Component Mixture Design: Good as Gold

Several thousand years ago, a jewelry maker discovered that the addition of copper to gold reduces the melt point of the resulting mixture. This led to a breakthrough in goldsmithing, because small decorations could be soldered with a copper-rich amalgam to a main element of pure gold. The copper blended in with no noticeable loss in luster of the finished piece.

Table 9-2. Alternative factorial design on fruit juices in terms of percentage

Std Order	A: Lemonade (%)	B: Apple Juice (%)	Total (%)
1	0	0	0
2	100	0	100
3	0	100	100
4	100	100	200

MIXTURE DESIGN

Table 9-3. A mixture experiment on copper and gold

Blend	A: Gold (wt %)	B: Copper (wt %)	Meltpoint (deg C)
Pure	100	0	1039, 1047
Binary	50	50	918, 922
Pure	0	100	1073, 1074

Table 9-3 lays out a simple mixture experiment aimed at quantifying this metallurgical phenomenon. This is a fully replicated design on the pure metals plus the binary (50/50) blend. To do a proper replication, each blend must be reformulated, not just retested for melt point. The latter approach would cause error to be underestimated, because the only variation would be that due to testing, not the entire process. Also, the run order must be randomized. Don't do the same formulation twice in a row, because you will more than likely get results that reflect less than the normal process variation.

Notice the depression in melt point in the blend. This is desirable and therefore an example of "synergistic" behavior. It can be modeled with the following equation, called a "mixture model":

$$\text{Melt point} = 1043.0A + 1073.5B - 553.2AB$$

where components A and B are expressed in proportional scale (0 to 1). This second-order model, developed by Henri Scheffé specifically for mixtures, is easy to interpret. The coefficients for the main effects are the responses for the purest "blends" for A and B. The negative coefficient on the interaction term (AB) indicates that a combination of the two components produces a response that's less that what you would expect from linear blending. This unexpected interaction becomes more obvious in the response surface graph shown in Figure 9-1.

> ### *EVERYTHING YOU ALWAYS WANT TO KNOW ABOUT INTERACTION (BUT WERE AFRAID TO ASK)*
>
> There are two ways that factors can interact: positively ("synergism") or negatively ("antagonism"). When synergism exists, you get a greater response than what you'd expect from simply adding the effects of each factor. In this case, adding one to one gives you more than two. For example, if you combined two firecrackers and got an explosion like an atomic bomb, you would be the beneficiary (victim?) of synergism. On the other hand, when antagonism occurs, you get less than what you'd expect from the combination of factors. As an analogy, one of the authors observes antagonism between his two similarly aged sons, and two similarly aged daughters. These are classic cases of sibling rivalry. A child may act very positively on his or her own, but in the presence of a sibling, they compete in a negative way for parental attention. Although antagonism is the norm, siblings can act synergistically, such as when they happily play together.

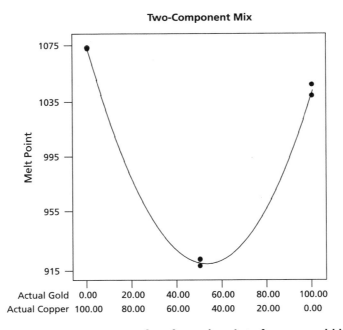

Figure 9-1. Response surface for melt point of copper-gold blend

For purposes of illustration, we've shown a simplified view of the actual behavior of copper and gold mixtures. The predicted value for equal amounts of gold and copper is:

$$\text{Melt point} = 1043.0 (0.5) + 1073.5 (0.5) - 553.2 (0.5 * 0.5)$$
$$= (1043.0 + 1073.5)/2 - 553.2/4 = 1058.25 - 138.3 = 919.95$$

The equation predicts a deflection of 138.3 degrees C from the expected melt point of 1058.25. Statistical analysis of the overall model and the interaction itself, shown in Table 9-4, reveals a significant fit.

Mixture problems require special treatment for computing probabilities. Specifically, the main effects (A and B in this case) cannot be independently estimated. For example, if you specify 40 percent of gold, then the copper must be 60 percent to make up the difference to the total of 100 percent. Therefore, the main effects for the mixture are collected in one test labelled "linear" in the ANOVA. Only one degree of freedom is available because the levels of gold and copper are completely interdependent—when the proportion of one component changes, the proportion of the other one does also. No matter how many components you include in a mixture design, the last one will always be fixed after setting the levels on all the others.

The ANOVA in this case lists "pure error" rather than "residual," because the estimate for error comes from replicated blends. Each pair of results contributes one degree of freedom for pure error, so there are three degrees of freedom in all.

Table 9-4. ANOVA for gold-copper mixture experiment

Source	Sum of Squares	Df	Mean Square	F Value	Prob >F
Model	26432.77	2	13216.39	969.30	< 0.0001
Linear	930.25	1	930.25	68.23	0.0037
AB	25502.52	1	25502.52	1870.37	< 0.0001
Pure Error	40.91	3	13.64		
Cor Total	26473.68	5			

> ### WORTH ITS WEIGHT IN GOLD?
>
> An ancient king suspected that his goldsmith mixed some silver into a supposedly pure gold crown. He asked the famous mathematician Archimedes to investigate. Archimedes performed the following experiment:
>
> 1. Create a bar of pure gold with the same weight as the crown.
> 2. Put the gold in a bath tub. Measure the volume of water spilled.
> 3. Do the same with the crown.
> 4. Compare the volumes.
>
> Archimedes knew that silver would be less dense than gold. Therefore, upon finding that the volume of the crown exceeded the volume of an equal weight of gold, he knew that the crown contained silver. According to legend, Archimedes then ran naked from his bath into the street shouting "Eureka!," which in Greek means "I have found it."
>
> The principles of mixture design can be put to work in this case. Gold and silver have densities of 10.2 and 5.5 troy ounces per cubic inch, respectively. Assume that no interactions exist between silver and gold in regard to density. We can then apply a linear mixture model to predict weight (in ounces) of one cubic inch (enough to make a crown?) as a function of proportional volume for gold (A) versus silver (B).
>
> Weight = 10.2 A + 5.5 B
>
> Notice that the coefficients of the model are simply the densities of the pure metals. The weight of a blend of half gold and half silver is calculated as follows:
>
> Weight = 10.2 (0.5) + 5.5 (0.5) = 7.85 troy ounces per cubic inch

Three-Component Design: Teeny Beany Experiment

Formulators usually experiment on more than two components. For example, let's look at a mixture design on three flavors of small jelly candies called "teeny beanies." Table 9-5 shows the blends and resulting taste ratings. The rating system, used earlier in Section 3 for popcorn experiments, goes from 0 (worst) to 10 (best). Each blend was rated by a panel, but in a different random order for each taster. The ratings were then averaged. Several of the blends were replicated to provide estimates of pure error.

This design is called a "simplex centroid" because it includes the blend with equal proportions of all the components. The centroid falls at the center of

MIXTURE DESIGN

Table 9-5. Teeny beany mixture design (three-component)

Blend	A: Apple %	B: Cinnamon %	C: Lemon %	Taste Rating
Pure	100.00	0.00	0.00	5.1, 5.2
Pure	0.00	100.00	0.00	6.5, 7.0
Pure	0.00	0.00	100.00	4.0, 4.5
Binary	50.00	50.00	0.00	6.9
Binary	50.00	0.00	50.00	2.8
Binary	0.00	50.00	50.00	3.5
Centroid	33.33	33.33	33.33	4.2, 4.3

the mixture space, which forms a simplex—a geometric term for a figure with one more vertex than the number of dimensions. For three components the simplex is an equilateral triangle. The addition of a fourth component creates a tetrahedron, which looks like a pyramid but with only three sides. Because only two-dimensional space can be represented on a piece of paper or computer screen, response data for mixtures of three or more components are displayed on triangular graphs such as the one shown in Figure 9-2.

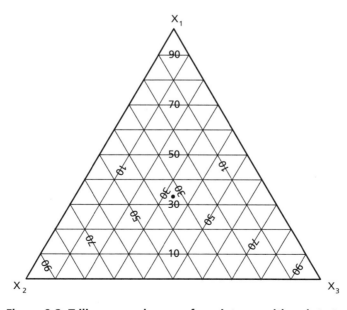

Figure 9-2. Trilinear graph paper for mixtures with point at centroid

Notice that the grid lines increase in value as you move from any of the three sides toward the opposing "vertex." The markings on this graph are in terms of percentage, so each vertex represents 100% of the labeled component. The unique feature of this trilinear graph paper is that only two components need to be specified. At this point the third component is fixed. For example, the combination of X_1 at 33.3% (1/3rd of the way up from the bottom) plus X_2 at 33.3% (1/3rd of the distance from the right side to lower left vertex) is all that's required to locate the centroid (shown on the graph). You can then read the value of the remaining component (X_3), which must be 33.3% to bring the total to approximately 100%.

Figure 9-3 shows a contour map fitted to the taste responses for the various blends of teeny beanies.

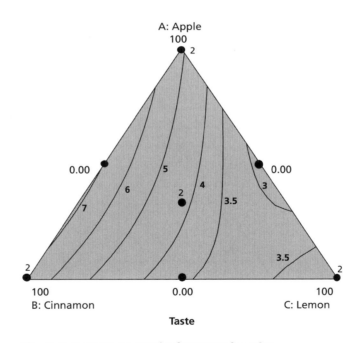

Figure 9-3. Contour graphs for teeny beanies

Notice the curves in the contours. This behavior is modeled by the following second-order Scheffé polynomial:

$$\text{Taste} = 5.14A + 6.74B + 4.24C + 4.12AB - 7.28AC - 7.68BC$$

Analysis of variance (not shown) indicates that this model is highly significant. Component B, cinnamon, exhibits the highest coefficient for the main effects. Thus, one can conclude that the best pure teeny beanie is cinnamon. Component C, the lemon flavor, was least preferred. Looking at the coefficients for the second-order terms, you can see by the positive coefficient that only the AB combination (apple-cinnamon) was rated favorably. The tasters down-rated combinations of apple-lemon (AC) and lemon-cinnamon (BC). These synergisms and antagonisms are manifested by upward and downward curves along the edges of the 3-D response surface shown in Figure 9-4.

This concludes our overview of mixture design, but we've only scratched the surface of this DOE specialty. For more detail, see the referenced text by Cornell.

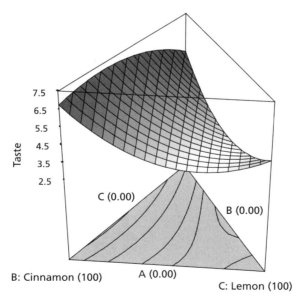

Figure 9-4. 3-D Response surface for taste of teeny beanies

CHAPTER

10
SOLUTIONS TO PRACTICE PROBLEMS

PROBLEM 1-1

The answers for the mean ($\bar{Y}$), standard error (SE), and approximate 95% confidence interval (CI), rounded to nearest tenth, are:

$$\bar{Y} = 58.0$$
$$SE = 1.1$$
$$CI \cong 58 \pm 2\,(1.1) = 58 \pm 2.2$$

A MORE ACCURATE SOLUTION THAT ILLUSTRATES HOW TO USE THE T-TABLE

The rule of thumb of two times standard error is slightly off in this case, because there are only 25 datapoints, which falls below the suggested minimum of 30. For a more accurate value of t, refer to the t-table in the Appendix. The objective of 95% or greater confidence is achieved by keeping the risk below 5%, or 0.05 on a scale of 0 to 1. Select this column of probability, which represents the proportion of area under the curve that falls outside of plus or minus t. (Notice that the drawing of the distribution, placed above the table, shows the two tails shaded.) The next step is to determine the degrees of freedom (df) so you know which row to use in the t-table. In this case we started with 25 datapoints, but used 1 degree of freedom to estimate the mean, so 24 degrees of freedom remain to estimate the variance. Looking across from the row to column 0.05, you will find a critical t-value of 2.064. Substituting this value into the formula for confidence interval produces the following result:

> CI = 58 ± 2.067(1.1) = 58 ± 2.27
>
> This outcome is not much different than the one obtained by using an approximate value for t.

PROBLEM 1-2

In this before-and-after case, the sample size (n) remains the same at 10, and the probability segments stay at 10% (100/10). As before, the lowest weight will be plotted at 5%, which is the midpoint of the first segment. Table 10-1 shows this combination and all the remaining ones for the original weight of the high school graduates, sorted from low to high. (Don't forget to do the sorting!)

Now all we need to do is plot the weights on the x-axis of the probability paper and the cumulative probabilities on the y-axis. See the resulting plot in Figure 10-1, which is a recycled version of Figure 1-8, but with 40 pounds dropped off each tic mark on the x-axis. (These guys really put on some weight in 25 years!)

Table 10-1. Values to plot on probability paper

Point	Weight	Cumulative Probability
1	130	5%
2	139	15%
3	147	25%
4	148	35%
5	153	45%
6	155	55%
7	161	65%
8	167	75%
9	178	85%
10	190	95%

SOLUTIONS TO PRACTICE PROBLEMS

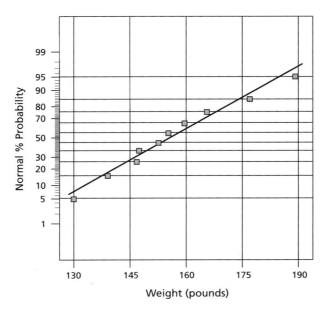

Figure 10-1. Normal plot of weights at time of graduation

The points on this plot generally line up, with no gross deviations. Therefore it's safe to say that this sample of individuals is most likely normal.

PROBLEM 2-1

For answers to the bowling case, see the "One Factor Tutorial" file, which comes with the DOE software that accompanies the book. Refer to the program instructions for details on installation and tutorials.

PROBLEM 2-2

The F-test on the second dice toss gives these results:

$$F = 0.21$$
$$\text{Prob} > F = 0.89$$

The F-ratio is again low and the associated probability high that any observed differences occurred due to chance. (Remember that the F-value must exceed 1 to have any hope for significance.) Figure 10-2 (software generated) shows the effects plot for the second toss of the colored dice. As you'd expect from the insignificant F-test, the LSD bars overlap, indicating no significant pairwise differences at the 95% confidence level. Therefore, there is no reason to dispute the earlier conclusion, from the first toss, that color does not significantly affect the outcome.

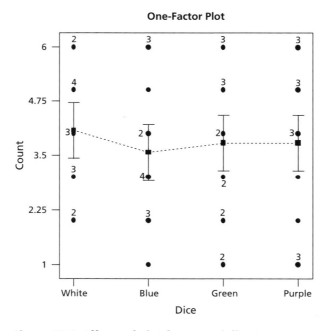

Figure 10-2. Effects of plot for second dice toss

SOLUTIONS TO PRACTICE PROBLEMS

PROBLEM 2-3

Statistical software produced the F-test results shown below (for the four lots of incoming shipments):

$$F = 21.24$$
$$\text{Prob} > F = {<}0.0001$$

The probability of getting an F-ratio this high due to chance is less than 1 out of 10,000. In other words, you can state with more than 99.99% confidence that at least one difference is significant. As you can see from Figure 10-3, the last shipment (#4) is much higher than the rest. The least significant difference bars do not overlap. In fact, if you look closely at sequential LSD bars, you can see a significant upward trend. Therefore, the delivery person's suspicions are correct—things have changed, and the supplier needs to be brought under control.

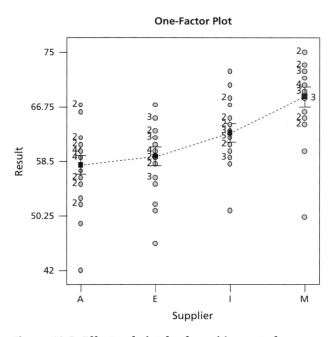

Figure 10-3. Effects of plot for four shipments from a supplier

> ### *A LESS PRECISE SOLUTION THAT ILLUSTRATES HOW TO USE F-TABLES (AND MAKES YOU GRATEFUL THAT STATISTICAL SOFTWARE NOW CAN DO ALL THIS FOR YOU!)*
>
> As a postscript to this solution, let's cross-check the reported probability for the F-value of 21.24. We can refer to one of the F-tables in the Appendix to make this determination. The drawings above the F-tables give you a feeling for the shape of the F-distribution. The shaded area represents the stated probability (10%, 5%, 1%, or 0.1%) of getting a value at or above the critical F. The table for 0.1% provides a level of probability equivalent to 0.001, reported on a scale of 0 to 1. The numerator of the F-test provides an estimate of the variance between the 4 shipment means. One degree of freedom is used to estimate the overall average of all shipments. That leaves 3 degrees of freedom for the numerator (df_{num}). The denominator of the F-test comes from the pooled variance within shipments. We collected 25 samples from each shipment for a total of 100 (4×25), but 4 degrees of freedom must be used to estimate the 4 shipment means, which leaves 96 degrees of freedom for the denominator (df_{den}). Now to look up the critical value of F for 0.1%, go to column 3 and row 96 or higher. The next highest row is for 120 degrees of freedom, which provides a conservative value of 5.781. The actual F of 21.24 exceeds this critical value, so you can conclude that it is significant at the 0.1% probability value (for risk)—which provides you with 99.99% confidence, as stated earlier.

PROBLEM 2-4

The ANOVA for the fabric test is shown in Table 10-2.

The Prob > F of 0.0016 falls far below the suggested cut-off of 0.05; thus, you can conclude that the outcome is significant. Notice the removal of the sum

Table 10-2. ANOVA for fabric test (blocked)

Source	Sum of Squares	Df	Mean Square	F Value	Prob >F
Block	30.47	8	3.81		
Model	1.13	1	1.13	21.95	0.0016
A	1.13	1	1.13	21.95	0.0016
Residual	0.41	8	0.051		
Cor Total	32.01	17			

SOLUTIONS TO PRACTICE PROBLEMS

of squares due to the blocks (the various subjects who wore the fabric). If this experiment had been done unblocked—where each subject got pants made only from one fabric—this variation (30.47 sum of squares) would go into the residual, inflating the mean square (MS) residual many-fold:

$$MS_{\text{Resid-Unblocked}} = \frac{SS_{\text{Block}} + SS_{\text{Resid-Blocked}}}{DF_{\text{Block}} + DF_{\text{Resid-Blocked}}} = \frac{30.47 + 0.41}{8 + 8} = \frac{30.88}{16} = 1.93$$

The F-value would then be less than 1 (1.13/1.93), which is not significant. The fabric effect is overwhelmed by the increased noise. Therefore, the decision to block the fabric test by subject proved to be the key to success.

Figure 10-4 displays the effect plot, showing only the mean wear for each fabric (actual responses from design points turned off). The least-significance difference bars do not overlap. Thus, one can say with more than 95% confidence that the new fabric wears better.

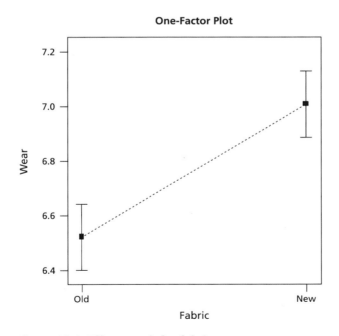

Figure 10-4. Effect graph for fabric test

DOE SIMPLIFIED

PROBLEM 3-1

For answers to the reactor case, see the "Factorial Tutorial" file, which comes with the DOE software that accompanies this book. Refer to the program instructions for details on installation and tutorials.

PROBLEM 3-2

The half-normal plot (Figure 10-5) reveals a family of effects: C, A, and the interaction AC. The other four effects (B, AC, BC, ABC) comprise the normal scatter of points near zero. These are deliberately left unlabeled as an indication of their insignificance.

The analysis of variance (Table 10-3) confirms the significance of the A, C, and AC effects—the vital few. All of them exhibit probability values below the recommended cutoff of 0.05. These three effects are pooled for the model, which comes out highly significant (Prob > F of 0.0076).

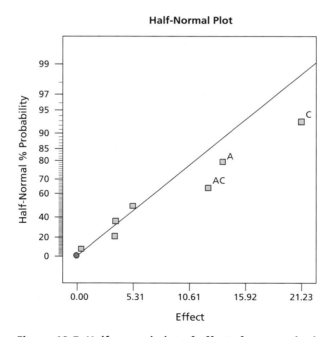

Figure 10-5. Half-normal plot of effects from car-shade experiment

SOLUTIONS TO PRACTICE PROBLEMS

Table 10-3. ANOVA for car-shade experiment

Source	Sum of Squares	Df	Mean Square	F Value	Prob >F
Model	1589.26	3	529.75	19.40	0.0076
A	379.50	1	379.50	13.90	0.0203
C	901.00	1	901.00	33.00	0.0046
AC	308.76	1	308.76	11.31	0.0282
Residual	109.22	4	27.31		
Cor Total	1698.49	7			

The statistics for the residual line on the ANOVA come from the pooling of the other four effects—the trivial many. The mean square of the residual provides an estimate of error that is used for the denominator of the F-values. Before moving ahead to the effect graphs, always check the residuals for normality (see Figure 10-6). The residuals do not show any dramatic nonlinearity, so it's OK to proceed by graphing the significant effects.

> ### THE LAST AND PERHAPS LEAST INTERESTING LINE OF THE STANDARD ANOVA TABLE
>
> The last line of the ANOVA tallies up all the sum of squares and degrees of freedom (DF). The label "Cor Total" indicates that this is the total variance corrected for the mean, which provides a baseline for statistical analysis.

Not surprisingly, the shaded location came out consistently coolest (factor C at plus level). You can easily pick this out from the raw response data shown in Table 3-9, but it helps to see a plot. Because factor C participates in the significant effect AC, it's best to view the interaction plot (see Figure 10-7), rather than the main effect plot. The flat line at the bottom of the plot indicates that in the shade (C+) it does not matter whether the window cover is white (A–) or shiny (A+). However, in the sun-drenched open lot (C–), the shiny side out (A+) worked best.

In fact, the overlapping LSD bars at the right of the plot show that the location of the car (factor C) makes no significant difference with the shiny cover. Therefore, there's no advantage to parking far away under the shade tree.

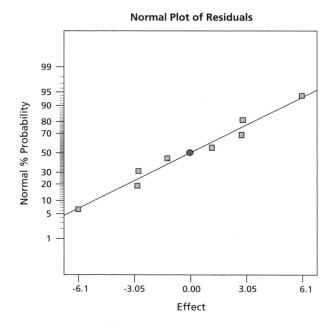

Figure 10-6. Normal plot of residuals

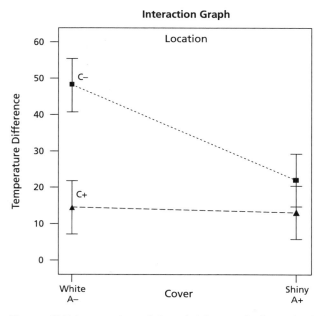

Figure 10-7. Interaction of A and C (sunny [C–] vs. shaded [C+])

SOLUTIONS TO PRACTICE PROBLEMS

PROBLEM 4-1

The half-normal table of effects (Figure 10-8) shows that the main effect of A (color) made the biggest impact on response rate for the postcard.

The ANOVA (not shown) supports the significance of A. However, the residual plots shown in Figure 10-9 look abnormal. Notice the increasing variation in the residuals versus predicted plot at the right. This is characteristic of a response that comes from a count, the variation of which can be expected to increase as the level goes up.

According to Table 4-4, a count is beneficially transformed by the square root. Figure 10-10 shows the plot of effects in this new response metric. Factor A remains the largest effect, but it's now joined by factor C (thickness) and moderated by factor B (size) in the form of interaction AB. The main effect of factor B also looks significant, with a small but appreciable gap between it and the line of remaining (deliberately unlabeled) effects near zero.

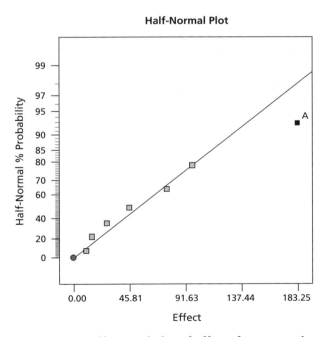

Figure 10-8. Half-normal plot of effects for postcard experiment

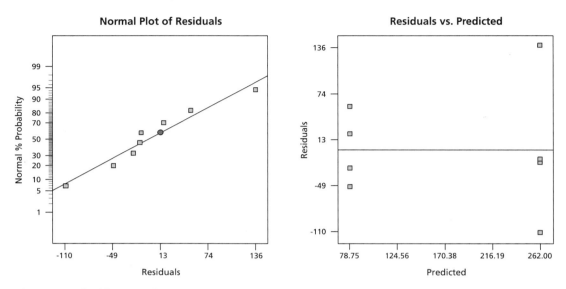

Figure 10-9 (a, b). Normal plots on normal paper (left) and predicted level (right)

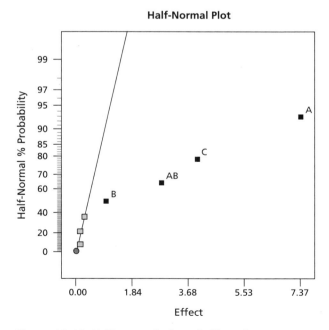

Figure 10-10. Half-normal plot of effects in square-root scale for postcard experiment

SOLUTIONS TO PRACTICE PROBLEMS

The ANOVA (not shown) supports the significance of the overall model and all chosen effects (including B). Furthermore, the residual plots shown in Figure 10-11 support statistical normality: straight line on the left (normal plot of residuals), and scatter on the right (residual versus predicted level).

The main effect of C is shown in Figure 10-12. This effect graph, and the next, are displayed in square root scale, which made the response linear. Regardless of scale, it's clear that the audience preferred thicker postcards. Perhaps the thinner stock did not fare as well in-transit through the bulk mail process (you know—bend, fold, spindle, and mutilate).

Figure 10-13 (p. 169) shows the interaction plot for AB. Surprisingly, the highly technical audience responded better to two colors (A–), saving many thousands in needless printing cost. Apparently, the fancier the piece, the more it turns off the techies: they want only the unembellished facts.

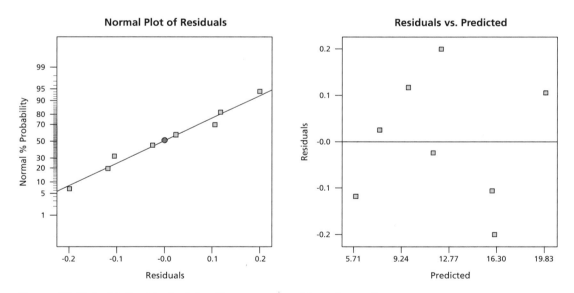

Figure 10-11 (a, b). Residual plots after square root transformation

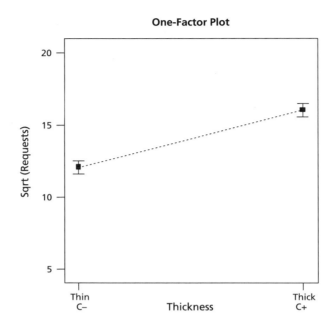

Figure 10-12. Main effect factor C (thickness)

The impact of color is modified by the size of the card (factor B). When the card is small (B−), the negative effect of color is lessened. A result like this would never be revealed by simple one-factor-at-a-time experimentation.

Putting all three factors together, the cube plot shown in Figure 10-14 reveals the ideal combination for the postcard at the upper, left, back corner: two-color (A−), large size (B+), and thick stock (C+). Remember to reverse the transformation before presenting your predictions. The cube at the right shows the squared outcomes, which convert the response back to original units. Not that it really matters, because the absolute numbers are only of value on a relative scale, but the maximum predicted response is 393.

The ideal corner of the response cube produces a medium level of cost, 10 cents per postcard, as shown on Figure 10-15 (p. 170). The big savings comes by avoiding the cost of four-color printing (A+), which would add four cents per postcard.

In the end, the market researchers saved a great deal of money by avoiding the more expensive configuration, while generating maximal response from

SOLUTIONS TO PRACTICE PROBLEMS

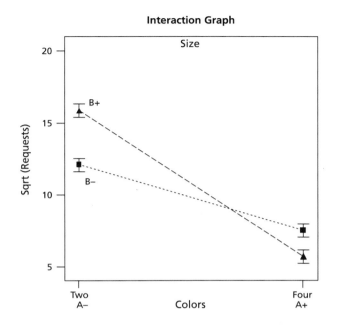

Figure 10-13. Interaction of A (color) and B (small [B–] versus large [B+] card)

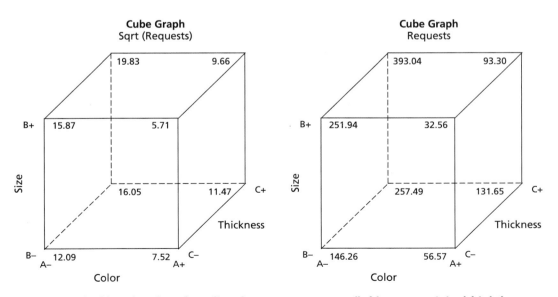

Figure 10-14 (a, b). Cube plot of predicted count–square root (left) versus original (right)

DOE SIMPLIFIED

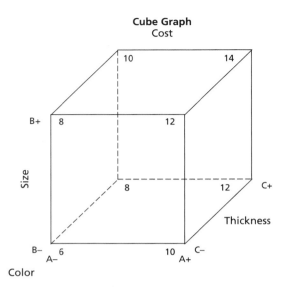

Figure 10-15. Costs (in cents per piece) of postcards made to varying specifications

the postcard. Does it make you a bit nervous to see marketing people using DOE to persuade you to buy their products?

PROBLEM 5-1

Figure 10-16 shows the plot of effects for the molding case. As expected, an interaction did occur between booster pressure (C) and moisture (D). These two main effects also show significance. The surprise was the BC interaction. B alone is not significant, but B is picked to support the interaction, thus maintaining model hierarchy.

The ANOVA (not shown) does not contradict any of the above conclusions. The variance due to blocks, removed before doing the F-test on the model, was relatively minor. Correlation of shrinkage with machine line was practically nonexistent, as seen in Figure 10-17 (p. 172). It can be concluded that blocking made no difference in the outcome. However, it was a prudent insurance policy against potential variation.

SOLUTIONS TO PRACTICE PROBLEMS

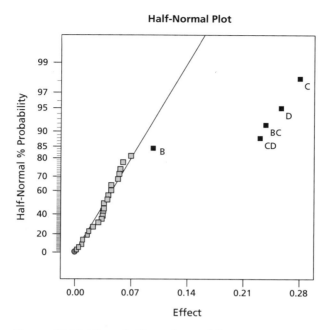

Figure 10-16. Plot of effects for molding case

The interaction plot for BC is shown in Figure 10-18. The best combination for minimum shrinkage with maximum throughput is B– (low cycle time) and C– (low booster pressure).

The DOE on the molding operation revealed another significant interaction: CD (Figure 10-19, p. 173). Notice that C (booster pressure) has an effect only when D (moisture) is high. Clearly, if you want to make shrinkage robust to booster pressure, it's best to keep moisture low (D–).

Before going any further, remember to check the alias structure (see Appendix 2-4). You will find that none of the significant effects is aliased with anything less than a three-factor interaction. Therefore, it's relatively safe to maintain the conclusions reached above. However, it's always advisable to do confirmation runs.

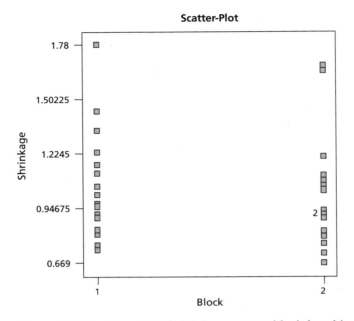

Figure 10-17. Scatter-plot of shrinkage versus block (machine line)

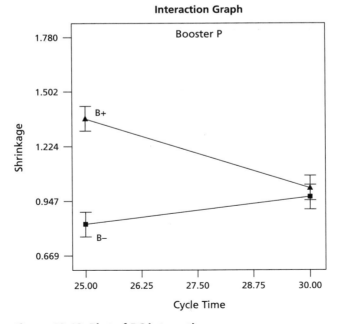

Figure 10-18. Plot of BC interaction

SOLUTIONS TO PRACTICE PROBLEMS

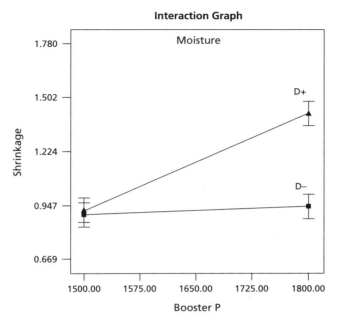

Figure 10-19. Plot of CD interaction

PROBLEM 6-1

The plot of effects shown in Figure 10-20 reveals several outstanding effects of skate configuration on the resulting times around the track.

The ANOVA verifies that something significant happened. The cube plot in Figure 10-21 tells the apparent story. The contrast from worst time (210.5 seconds) to best time (165.5 seconds) represents a very noticeable improvement of 45 seconds. Notice on the cube plot that changing the bearing from the old (B–) to the new material (B+) caused a decrease in time. This makes sense, as does the effect of wheels. You can see that going to the hard wheels (E–) decreases the time. These two changes were put into effect immediately. Factor G, neon lighting, was put in as a factor because the generator needed to power the lights creates some drag. Notice that the times do increase when the lights go on (G+). However, what's listed as G could be an aliased effect.

DOE SIMPLIFIED

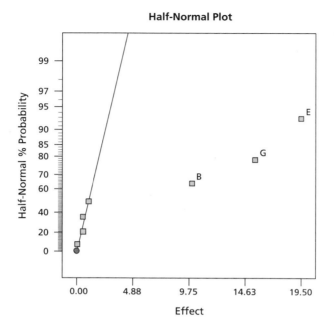

Figure 10-20. Plot of effects for in-line skating experiment

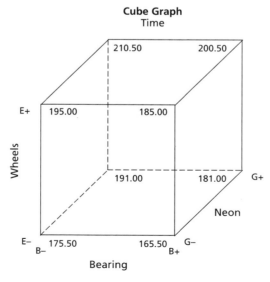

Figure 10-21. Cube plot of significant effects on skating (time in seconds)

SOLUTIONS TO PRACTICE PROBLEMS

Referring to the alias structure shown on Table 6-3 (p. 113), we see that: G = G + AF + BE + CD. Perhaps G is really the interaction BE—which makes sense, because both parents already appear significant. Move on to the next problem for the rest of the story.

PROBLEM 6-2

As suspected by the experimenter, the fold-over revealed he was spinning his wheels over the effect of neon lighting (G). The plot of effects in Figure 10-22 shows the interaction AF instead. This seemed odd, because neither of the parent effects came out significant, but it couldn't be ruled out on this basis alone.

The ANOVA (not shown) looks good, with no obvious non-normality in the residuals. However, the experimenter could not come up with a plausible reason for an interaction involving the heel pad (A) and the wheel covers (F), so he inspected the aliased effects: [AF] = AF + BE + CD. The interaction of C (gloves on or off) and D (helmet logo back or front) is just as implausible, so it was rejected. The only other possibility—interaction BE—made sense, not only physically, but because it derives from parent effects already in the predictive

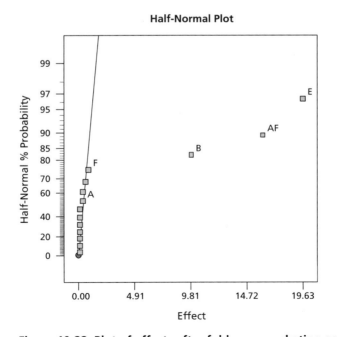

Figure 10-22. Plot of effects after fold-over on skating equipment

DOE SIMPLIFIED

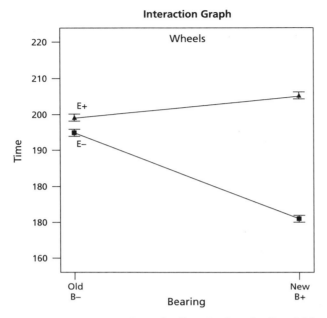

Figure 10-23. Interaction of B (bearing) and E (hard [–] vs soft [+] wheel)

model. This is shown in Figure 10-23, which shows consistently poorer times with the soft wheels (E+). When coupled with the hard wheels (E–), the new bearings made of high-tech alloy (B+) gave a big boost to the skater, who now speeded around the track in record time (at least for him). Confirmation runs provided support for the interaction of bearings and wheels. The two-part DOE proved to be a big success except for one thing: how to stop without crashing.

PROBLEM 7-1

For answers to the battery case, see the "General Factorial Tutorial" file, which comes with the DOE software that accompanies this book. Refer to the program instructions for details on installation and tutorials.

PROBLEM 7-2

The ANOVA shown in Table 10-4 indicates that route (A) and departure time (B) significantly affected commuting time. Using a probability value (Prob>F) of 0.05 as a cutoff, it can be concluded these two factors (AB) interact.

SOLUTIONS TO PRACTICE PROBLEMS

Table 10-4. ANOVA for driving DOE

Source	Sum of Squares	Df	Mean Square	F Value	Prob >F
Model	37.52	5	7.50	10.79	0.0059
A	5.86	2	2.93	4.21	0.0720
B	23.69	1	23.69	34.05	0.0011
AB	7.97	2	3.99	5.73	0.0406
Residual	4.17	6	0.70		
Lack of Fit	0.000	0			
Pure Error	4.17	6	0.70		
Cor Total	41.70	11			

On the interaction plot (Figure 10-24), you can see that the central route gave the most consistent results. The commute times for the alternative routes—south, and to a lesser extent, north—increase significantly when departing late. When departing early, any of the three routes can be chosen. Perhaps with

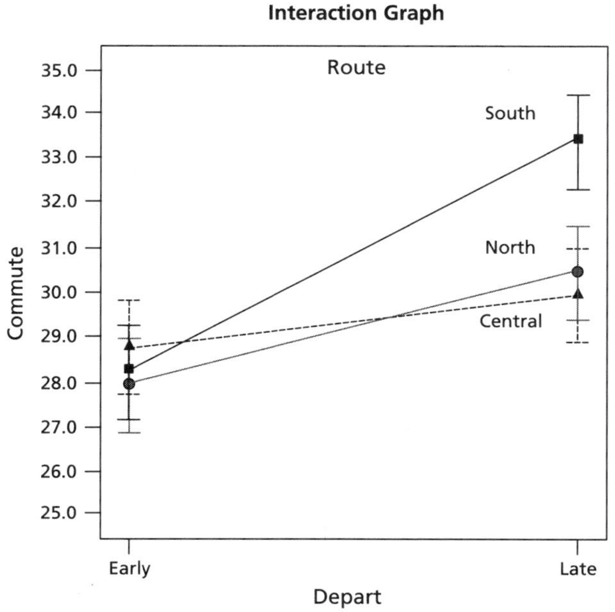

Figure 10-24. Interaction graph for driving DOE

more iterations, the commuter will recognize a one-minute or so advantage to going north or south, but the latter route must definitely be avoided if late.

> ### *A MINUTE SAVED IS A MINUTE EARNED!*
>
> Is it worth worrying about a possible one-minute savings? Look at it this way: Over the course of a year a person goes to work as many as 250 days. Let's make a big assumption and say that a minute can be saved on the return home also. So now we're talking about 2 minutes per day saved. Multiplied by 250 this time becomes 500 minutes, or a bit over 8 hours. This represents an entire day saved! Too bad it's spread so thin.

CHAPTER 11
PRACTICE EXPERIMENTS

PRACTICE EXPERIMENT #1: BREAKING PAPER CLIPS

It's easy to find various brands, types (for example, virgin or recycled, or rough-edged versus smooth), and sizes of paper clips. These can be tested for strength via a simple comparative experiment, such as those illustrated in Chapter 2. (You might also find this to be a good way to relieve stress!) The procedure shown below specifies big versus little paper clips. It's intended for a group of people, with each individual doing the test. The person-to-person variation should be treated as blocks and removed from the analysis. Also, the test goes a lot quicker and easier this way, instead of doing it all yourself. The procedure is as follows:

1. Randomly select one big and one regular-sized paper clip. Flip a coin to randomly choose the first clip to break: heads = big, tails = regular.
2. Gently pull each clip apart with the big loop on the right. Use the drawing in Figure 11-1 as a template. The angle affects performance, so be precise.
3. As pictured in Figure 11-1, move the smaller loop of the clip to the edge of your desk. The bigger loop should now project beyond the edge.
4. Hold the small loop down firmly with your left thumb. Grasp the big loop between right thumb and forefinger. Bend the big loop straight up and back. Continue bending the big loop back and forth until it breaks. Record the count for each clip. (Each back and forth movement counts as two bends.)

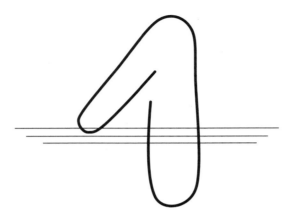

Figure 11-1. Paper clip positioned at edge of desk

Tabulate the results from the entire group. Do a statistical analysis of the data in a manner similar to that outlined in Chapter 2, Problem 4 (wear tests on old versus new fabric). Is one paper clip significantly stronger than the other? (Suggestion: use the software provided with this book. Set up a one-factor design similar to that shown in the tutorial that comes with the program. Be sure to identify this as a blocked design. Treat each person as a block. The ANOVA will then remove the person-to-person variance before doing the F-test on the effect of changing the type of paper clip. If the test is significant, make an effects plot.)

PRACTICE EXPERIMENTS

PRACTICE EXPERIMENT #2: HAND-EYE COORDINATION

This is a simple test of hand-eye coordination that nicely illustrates an interaction of two factors:

1. Hand, left ("L") versus right ("R").
2. Diameter of a target, small versus large.

Obviously, results will vary from one person to the next depending on which hand is dominant. The procedure specifies four combinations in a two-by-two factorial (see Figure 11-2 below).

We recommend you replicate the design at least twice (eight runs total) to add power for the statistical analysis. The procedure is as follows:

1. Draw two pairs of circles, centers equally distant, but with diameters that differ by a factor of two or more. Or copy the template shown below in Figure 11-3.
2. Randomly select either your left or right hand and one set of circles.

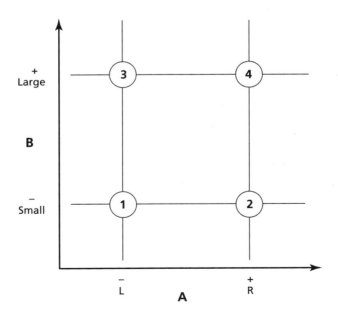

Figure 11-2. Design layout for hand-eye coordination test

181

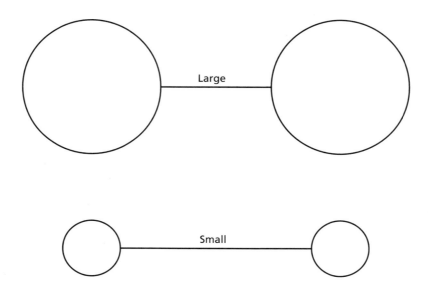

Figure 11-3. Template for hand-eye coordination test

3. Alternating between circles in a set, mark as many dots as you can in 10 seconds. Your score will consist of the number of paired, in-target cycles that you complete in 10 seconds. A cycle means a dot in both circles—back and forth once. We recommend that you count the cycles as you are performing the experiment, because with so many dots in a confined space, it's very hard to tell them apart.
4. When the time is up, subtract one from your count for each dot outside either circle. Record your corrected count as the score.
5. Repeat the above steps to complete all four combinations at least twice each, for a total of at least eight runs.

Analyze the results in the same manner as that outlined in Chapter 3. Do an ANOVA to see if anything is significant. Watch for an interaction of factors. (Suggestion: use the software provided with the book. Set up a factorial design, similar to the one you did for the tutorial that comes with the program, for two factors in four runs, with 2 replicates. Be sure to enter the data properly so the inputs match up with the outputs. Then do the analysis as outlined in the tutorial.)

PRACTICE EXPERIMENTS

OTHER FUN IDEAS FOR PRACTICE EXPERIMENTS

1. Ball in Funnel

This experiment is loosely based on Deming's funnel experiment. Time how long it takes a ball to spin through a funnel set at various heights. The ball can be fed through a tube. Vary the inclination and entry angle. Consider using different types of balls. Fasten the funnel so it's somewhat loose. You might then find that the effect of ball size depends on whether or not you hold the funnel—an interaction. Many more factors can be studied. Have a ball! (For details, see Bert Gunter, "Through a Funnel Slowly with Ball Bearing and Insight to Teach Experimental Design," *The American Statistician*, Vol. 47, Nov. 1993.)

2. Flight of the Balsa Buzzard

This is a fun DOE that anyone can do. Depending on your ambition, purchase 10 to 20 balsa airplanes at your local hobby shop. We suggest you test five factors: vertical stabilizer frontward or backward; the same for the horizontal stabilizer; wing position to the front or back; pilot in or out; and nose weight as is or increased. If you do test these five factors, try a half-fraction of a two-level factorial. For each configuration, make two flights. Input the mean distance and range as separate responses. Watch out—you may discover that certain factors cause increased variation in the flight path. (Statistician Roger Longbotham contributed this idea to the authors.)

3. Paper Airplanes

This experiment makes school teachers cringe. Students won't need much help. Let them apply their imagination to identify factors. Here are some ideas from grad students at North Carolina Tech: use multiple sheets, alter the design, change the width and length, increase launch height and/or angle. Desired responses are length and accuracy. (For details, see Sanjiv Sarin, "Teaching Taguchi's Approach to Parameter Design," *Quality Progress*, May 1997.)

4. Impact Craters:

Drop ball bearings (or marbles) of varying size into shallow containers filled with fine sand or granular sugar. Measure the diameter of the resulting crater. Try different drop heights and any other factors you come up with. Be prepared for some powerful interactions. If you do this with children, put some little toy dinosaurs in the sand. Count how many become extinct. (For details, see Bert Gunter, *Linking High School Math and Science Through Statistical Design of Experiments*, Macomb Intermediate School District, 1995, page 2-1. Also see the web site, www.macomb.k12.mi.us/math/web1.htm, on the Internet.)

APPENDICES

Appendix 1-1: Two-tailed t-Table

Probability points of the t-distribution with df degrees of freedom

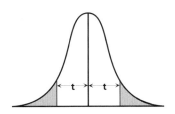

df	Two-tail area probability					
	0.2	0.1	0.05	0.01	0.005	0.001
1	3.078	6.314	12.706	63.657	127.321	636.619
2	1.886	2.920	4.303	9.925	14.089	31.599
3	1.638	2.353	3.182	5.841	7.453	12.924
4	1.533	2.132	2.776	4.604	5.598	8.610
5	1.476	2.015	2.571	4.032	4.773	6.869
6	1.440	1.943	2.447	3.707	4.317	5.959
7	1.415	1.895	2.365	3.499	4.029	5.408
8	1.397	1.860	2.306	3.355	3.833	5.041
9	1.383	1.833	2.262	3.250	3.690	4.781
10	1.372	1.812	2.228	3.169	3.581	4.587
11	1.363	1.796	2.201	3.106	3.497	4.437
12	1.356	1.782	2.179	3.055	3.428	4.318
13	1.350	1.771	2.160	3.012	3.372	4.221
14	1.345	1.761	2.145	2.977	3.326	4.140
15	1.341	1.753	2.131	2.947	3.286	4.073
16	1.337	1.746	2.120	2.921	3.252	4.015
17	1.333	1.740	2.110	2.898	3.222	3.965
18	1.330	1.734	2.101	2.878	3.197	3.922
19	1.328	1.729	2.093	2.861	3.174	3.883
20	1.325	1.725	2.086	2.845	3.153	3.850
21	1.323	1.721	2.080	2.831	3.135	3.819
22	1.321	1.717	2.074	2.819	3.119	3.792
23	1.319	1.714	2.069	2.807	3.104	3.768
24	1.318	1.711	2.064	2.797	3.091	3.745
25	1.316	1.708	2.060	2.787	3.078	3.725
26	1.315	1.706	2.056	2.779	3.067	3.707
27	1.314	1.703	2.052	2.771	3.057	3.690
28	1.313	1.701	2.048	2.763	3.047	3.674
29	1.311	1.699	2.045	2.756	3.038	3.659
30	1.310	1.697	2.042	2.750	3.030	3.646
40	1.303	1.684	2.021	2.704	2.971	3.551
60	1.296	1.671	2.000	2.660	2.915	3.460
120	1.289	1.658	1.980	2.617	2.860	3.373
1000	1.282	1.646	1.962	2.581	2.813	3.300
10000	1.282	1.645	1.960	2.576	2.808	3.290

Appendix 1-2: F-Table for 10%

Percentage points of the F-distribution:
upper 10% points

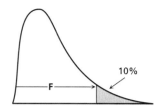

df_{den} \ df_{num}	1	2	3	4	5	6	7	8	9	10	15	20
1	39.863	49.500	53.593	55.833	57.240	58.204	58.906	59.439	59.858	60.195	61.220	61.740
2	8.526	9.000	9.162	9.243	9.293	9.326	9.349	9.367	9.381	9.392	9.425	9.441
3	5.538	5.462	5.391	5.343	5.309	5.285	5.266	5.252	5.240	5.230	5.200	5.184
4	4.545	4.325	4.191	4.107	4.051	4.010	3.979	3.955	3.936	3.920	3.870	3.844
5	4.060	3.780	3.619	3.520	3.453	3.405	3.368	3.339	3.316	3.297	3.238	3.207
6	3.776	3.463	3.289	3.181	3.108	3.055	3.014	2.983	2.958	2.937	2.871	2.836
7	3.589	3.257	3.074	2.961	2.883	2.827	2.785	2.752	2.725	2.703	2.632	2.595
8	3.458	3.113	2.924	2.806	2.726	2.668	2.624	2.589	2.561	2.538	2.464	2.425
9	3.360	3.006	2.813	2.693	2.611	2.551	2.505	2.469	2.440	2.416	2.340	2.298
10	3.285	2.924	2.728	2.605	2.522	2.461	2.414	2.377	2.347	2.323	2.244	2.201
11	3.225	2.860	2.660	2.536	2.451	2.389	2.342	2.304	2.274	2.248	2.167	2.123
12	3.177	2.807	2.606	2.480	2.394	2.331	2.283	2.245	2.214	2.188	2.105	2.060
13	3.136	2.763	2.560	2.434	2.347	2.283	2.234	2.195	2.164	2.138	2.053	2.007
14	3.102	2.726	2.522	2.395	2.307	2.243	2.193	2.154	2.122	2.095	2.010	1.962
15	3.073	2.695	2.490	2.361	2.273	2.208	2.158	2.119	2.086	2.059	1.972	1.924
16	3.048	2.668	2.462	2.333	2.244	2.178	2.128	2.088	2.055	2.028	1.940	1.891
17	3.026	2.645	2.437	2.308	2.218	2.152	2.102	2.061	2.028	2.001	1.912	1.862
18	3.007	2.624	2.416	2.286	2.196	2.130	2.079	2.038	2.005	1.977	1.887	1.837
19	2.990	2.606	2.397	2.266	2.176	2.109	2.058	2.017	1.984	1.956	1.865	1.814
20	2.975	2.589	2.380	2.249	2.158	2.091	2.040	1.999	1.965	1.937	1.845	1.794
21	2.961	2.575	2.365	2.233	2.142	2.075	2.023	1.982	1.948	1.920	1.827	1.776
22	2.949	2.561	2.351	2.219	2.128	2.060	2.008	1.967	1.933	1.904	1.811	1.759
23	2.937	2.549	2.339	2.207	2.115	2.047	1.995	1.953	1.919	1.890	1.796	1.744
24	2.927	2.538	2.327	2.195	2.103	2.035	1.983	1.941	1.906	1.877	1.783	1.730
25	2.918	2.528	2.317	2.184	2.092	2.024	1.971	1.929	1.895	1.866	1.771	1.718
26	2.909	2.519	2.307	2.174	2.082	2.014	1.961	1.919	1.884	1.855	1.760	1.706
27	2.901	2.511	2.299	2.165	2.073	2.005	1.952	1.909	1.874	1.845	1.749	1.695
28	2.894	2.503	2.291	2.157	2.064	1.996	1.943	1.900	1.865	1.836	1.740	1.685
29	2.887	2.495	2.283	2.149	2.057	1.988	1.935	1.892	1.857	1.827	1.731	1.676
30	2.881	2.489	2.276	2.142	2.049	1.980	1.927	1.884	1.849	1.819	1.722	1.667
40	2.835	2.440	2.226	2.091	1.997	1.927	1.873	1.829	1.793	1.763	1.662	1.605
60	2.791	2.393	2.177	2.041	1.946	1.875	1.819	1.775	1.738	1.707	1.603	1.543
120	2.748	2.347	2.130	1.992	1.896	1.824	1.767	1.722	1.684	1.652	1.545	1.482
100K	2.706	2.303	2.084	1.945	1.847	1.774	1.717	1.670	1.632	1.599	1.487	1.421

K (multiply this value by 1,000)

Appendix 1-3: F-Table for 5%

Percentage points of the F-distribution:
upper 5% points

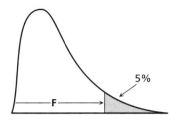

df_{den} \ df_{num}	1	2	3	4	5	6	7	8	9	10	15	20
1	161.45	199.50	215.71	224.58	230.16	233.99	236.77	238.88	240.54	241.88	245.95	248.01
2	18.513	19.000	19.164	19.247	19.296	19.330	19.353	19.371	19.385	19.396	19.429	19.446
3	10.128	9.552	9.277	9.117	9.013	8.941	8.887	8.845	8.812	8.786	8.703	8.660
4	7.709	6.944	6.591	6.388	6.256	6.163	6.094	6.041	5.999	5.964	5.858	5.803
5	6.608	5.786	5.409	5.192	5.050	4.950	4.876	4.818	4.772	4.735	4.619	4.558
6	5.987	5.143	4.757	4.534	4.387	4.284	4.207	4.147	4.099	4.060	3.938	3.874
7	5.591	4.737	4.347	4.120	3.972	3.866	3.787	3.726	3.677	3.637	3.511	3.445
8	5.318	4.459	4.066	3.838	3.687	3.581	3.500	3.438	3.388	3.347	3.218	3.150
9	5.117	4.256	3.863	3.633	3.482	3.374	3.293	3.230	3.179	3.137	3.006	2.936
10	4.965	4.103	3.708	3.478	3.326	3.217	3.135	3.072	3.020	2.978	2.845	2.774
11	4.844	3.982	3.587	3.357	3.204	3.095	3.012	2.948	2.896	2.854	2.719	2.646
12	4.747	3.885	3.490	3.259	3.106	2.996	2.913	2.849	2.796	2.753	2.617	2.544
13	4.667	3.806	3.411	3.179	3.025	2.915	2.832	2.767	2.714	2.671	2.533	2.459
14	4.600	3.739	3.344	3.112	2.958	2.848	2.764	2.699	2.646	2.602	2.463	2.388
15	4.543	3.682	3.287	3.056	2.901	2.790	2.707	2.641	2.588	2.544	2.403	2.328
16	4.494	3.634	3.239	3.007	2.852	2.741	2.657	2.591	2.538	2.494	2.352	2.276
17	4.451	3.592	3.197	2.965	2.810	2.699	2.614	2.548	2.494	2.450	2.308	2.230
18	4.414	3.555	3.160	2.928	2.773	2.661	2.577	2.510	2.456	2.412	2.269	2.191
19	4.381	3.522	3.127	2.895	2.740	2.628	2.544	2.477	2.423	2.378	2.234	2.155
20	4.351	3.493	3.098	2.866	2.711	2.599	2.514	2.447	2.393	2.348	2.203	2.124
21	4.325	3.467	3.072	2.840	2.685	2.573	2.488	2.420	2.366	2.321	2.176	2.096
22	4.301	3.443	3.049	2.817	2.661	2.549	2.464	2.397	2.342	2.297	2.151	2.071
23	4.279	3.422	3.028	2.796	2.640	2.528	2.442	2.375	2.320	2.275	2.128	2.048
24	4.260	3.403	3.009	2.776	2.621	2.508	2.423	2.355	2.300	2.255	2.108	2.027
25	4.242	3.385	2.991	2.759	2.603	2.490	2.405	2.337	2.282	2.236	2.089	2.007
26	4.225	3.369	2.975	2.743	2.587	2.474	2.388	2.321	2.265	2.220	2.072	1.990
27	4.210	3.354	2.960	2.728	2.572	2.459	2.373	2.305	2.250	2.204	2.056	1.974
28	4.196	3.340	2.947	2.714	2.558	2.445	2.359	2.291	2.236	2.190	2.041	1.959
29	4.183	3.328	2.934	2.701	2.545	2.432	2.346	2.278	2.223	2.177	2.027	1.945
30	4.171	3.316	2.922	2.690	2.534	2.421	2.334	2.266	2.211	2.165	2.015	1.932
40	4.085	3.232	2.839	2.606	2.449	2.336	2.249	2.180	2.124	2.077	1.924	1.839
60	4.001	3.150	2.758	2.525	2.368	2.254	2.167	2.097	2.040	1.993	1.836	1.748
120	3.920	3.072	2.680	2.447	2.290	2.175	2.087	2.016	1.959	1.910	1.750	1.659
100K	3.842	2.996	2.605	2.372	2.214	2.099	2.010	1.939	1.880	1.831	1.666	1.571

K *(multiply this value by 1,000)*

Appendix 1-4: F-Table for 1%

Percentage points of the F-distribution:
upper 1% points

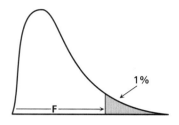

df_{den} \ df_{num}	1	2	3	4	5	6	7	8	9	10	15	20
1	4052.2	4999.5	5403.3	5624.6	5763.6	5859.0	5928.3	5981.1	6022.5	6055.8	6157.3	6208.7
2	98.503	99.000	99.166	99.249	99.299	99.333	99.356	99.374	99.388	99.399	99.433	99.449
3	34.116	30.817	29.457	28.710	28.237	27.911	27.672	27.489	27.345	27.229	26.872	26.690
4	21.198	18.000	16.694	15.977	15.522	15.207	14.976	14.799	14.659	14.546	14.198	14.020
5	16.258	13.274	12.060	11.392	10.967	10.672	10.456	10.289	10.158	10.051	9.722	9.553
6	13.745	10.925	9.780	9.148	8.746	8.466	8.260	8.102	7.976	7.874	7.559	7.396
7	12.246	9.547	8.451	7.847	7.460	7.191	6.993	6.840	6.719	6.620	6.314	6.155
8	11.259	8.649	7.591	7.006	6.632	6.371	6.178	6.029	5.911	5.814	5.515	5.359
9	10.561	8.022	6.992	6.422	6.057	5.802	5.613	5.467	5.351	5.257	4.962	4.808
10	10.044	7.559	6.552	5.994	5.636	5.386	5.200	5.057	4.942	4.849	4.558	4.405
11	9.646	7.206	6.217	5.668	5.316	5.069	4.886	4.744	4.632	4.539	4.251	4.099
12	9.330	6.927	5.953	5.412	5.064	4.821	4.640	4.499	4.388	4.296	4.010	3.858
13	9.074	6.701	5.739	5.205	4.862	4.620	4.441	4.302	4.191	4.100	3.815	3.665
14	8.862	6.515	5.564	5.035	4.695	4.456	4.278	4.140	4.030	3.939	3.656	3.505
15	8.683	6.359	5.417	4.893	4.556	4.318	4.142	4.004	3.895	3.805	3.522	3.372
16	8.531	6.226	5.292	4.773	4.437	4.202	4.026	3.890	3.780	3.691	3.409	3.259
17	8.400	6.112	5.185	4.669	4.336	4.102	3.927	3.791	3.682	3.593	3.312	3.162
18	8.285	6.013	5.092	4.579	4.248	4.015	3.841	3.705	3.597	3.508	3.227	3.077
19	8.185	5.926	5.010	4.500	4.171	3.939	3.765	3.631	3.523	3.434	3.153	3.003
20	8.096	5.849	4.938	4.431	4.103	3.871	3.699	3.564	3.457	3.368	3.088	2.938
21	8.017	5.780	4.874	4.369	4.042	3.812	3.640	3.506	3.398	3.310	3.030	2.880
22	7.945	5.719	4.817	4.313	3.988	3.758	3.587	3.453	3.346	3.258	2.978	2.827
23	7.881	5.664	4.765	4.264	3.939	3.710	3.539	3.406	3.299	3.211	2.931	2.781
24	7.823	5.614	4.718	4.218	3.895	3.667	3.496	3.363	3.256	3.168	2.889	2.738
25	7.770	5.568	4.675	4.177	3.855	3.627	3.457	3.324	3.217	3.129	2.850	2.699
26	7.721	5.526	4.637	4.140	3.818	3.591	3.421	3.288	3.182	3.094	2.815	2.664
27	7.677	5.488	4.601	4.106	3.785	3.558	3.388	3.256	3.149	3.062	2.783	2.632
28	7.636	5.453	4.568	4.074	3.754	3.528	3.358	3.226	3.120	3.032	2.753	2.602
29	7.598	5.420	4.538	4.045	3.725	3.499	3.330	3.198	3.092	3.005	2.726	2.574
30	7.562	5.390	4.510	4.018	3.699	3.473	3.304	3.173	3.067	2.979	2.700	2.549
40	7.314	5.179	4.313	3.828	3.514	3.291	3.124	2.993	2.888	2.801	2.522	2.369
60	7.077	4.977	4.126	3.649	3.339	3.119	2.953	2.823	2.718	2.632	2.352	2.198
120	6.851	4.787	3.949	3.480	3.174	2.956	2.792	2.663	2.559	2.472	2.192	2.035
100K	6.635	4.605	3.782	3.319	3.017	2.802	2.640	2.511	2.408	2.321	2.039	1.878

K (multiply this value by 1,000)

Appendix 1-5: F-Table for 0.1%

Percentage points of the F-distribution:
upper 0.1% points

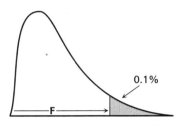

df_{den} \ df_{num}	1	2	3	4	5	6	7	8	9	10	15	20
1	405.2K	500.0K	540.4K	562.5K	576.4K	585.9K	592.9K	598.1K	602.3K	605.6K	615.8K	620.9K
2	998.50	999.00	999.17	999.25	999.30	999.33	999.36	999.37	999.39	999.40	999.43	999.45
3	167.03	148.50	141.11	137.10	134.58	132.85	131.58	130.62	129.86	129.25	127.37	126.42
4	74.137	61.246	56.177	53.436	51.712	50.525	49.658	48.996	48.475	48.053	46.761	46.100
5	47.181	37.122	33.202	31.085	29.752	28.834	28.163	27.649	27.244	26.917	25.911	25.395
6	35.507	27.000	23.703	21.924	20.803	20.030	19.463	19.030	18.688	18.411	17.559	17.120
7	29.245	21.689	18.772	17.198	16.206	15.521	15.019	14.634	14.330	14.083	13.324	12.932
8	25.415	18.494	15.829	14.392	13.485	12.858	12.398	12.046	11.767	11.540	10.841	10.480
9	22.857	16.387	13.902	12.560	11.714	11.128	10.698	10.368	10.107	9.894	9.238	8.898
10	21.040	14.905	12.553	11.283	10.481	9.926	9.517	9.204	8.956	8.754	8.129	7.804
11	19.687	13.812	11.561	10.346	9.578	9.047	8.655	8.355	8.116	7.922	7.321	7.008
12	18.643	12.974	10.804	9.633	8.892	8.379	8.001	7.710	7.480	7.292	6.709	6.405
13	17.815	12.313	10.209	9.073	8.354	7.856	7.489	7.206	6.982	6.799	6.231	5.934
14	17.143	11.779	9.729	8.622	7.922	7.436	7.077	6.802	6.583	6.404	5.848	5.557
15	16.587	11.339	9.335	8.253	7.567	7.092	6.741	6.471	6.256	6.081	5.535	5.248
16	16.120	10.971	9.006	7.944	7.272	6.805	6.460	6.195	5.984	5.812	5.274	4.992
17	15.722	10.658	8.727	7.683	7.022	6.562	6.223	5.962	5.754	5.584	5.054	4.775
18	15.379	10.390	8.487	7.459	6.808	6.355	6.021	5.763	5.558	5.390	4.866	4.590
19	15.081	10.157	8.280	7.265	6.622	6.175	5.845	5.590	5.388	5.222	4.704	4.430
20	14.819	9.953	8.098	7.096	6.461	6.019	5.692	5.440	5.239	5.075	4.562	4.290
21	14.587	9.772	7.938	6.947	6.318	5.881	5.557	5.308	5.109	4.946	4.437	4.167
22	14.380	9.612	7.796	6.814	6.191	5.758	5.438	5.190	4.993	4.832	4.326	4.058
23	14.195	9.469	7.669	6.696	6.078	5.649	5.331	5.085	4.890	4.730	4.227	3.961
24	14.028	9.339	7.554	6.589	5.977	5.550	5.235	4.991	4.797	4.638	4.139	3.873
25	13.877	9.223	7.451	6.493	5.885	5.462	5.148	4.906	4.713	4.555	4.059	3.794
26	13.739	9.116	7.357	6.406	5.802	5.381	5.070	4.829	4.637	4.480	3.986	3.723
27	13.613	9.019	7.272	6.326	5.726	5.308	4.998	4.759	4.568	4.412	3.920	3.658
28	13.498	8.931	7.193	6.253	5.656	5.241	4.933	4.695	4.505	4.349	3.859	3.598
29	13.391	8.849	7.121	6.186	5.593	5.179	4.873	4.636	4.447	4.292	3.804	3.543
30	13.293	8.773	7.054	6.125	5.534	5.122	4.817	4.581	4.393	4.239	3.753	3.493
40	12.609	8.251	6.595	5.698	5.128	4.731	4.436	4.207	4.024	3.874	3.400	3.145
60	11.973	7.768	6.171	5.307	4.757	4.372	4.086	3.865	3.687	3.541	3.078	2.827
120	11.380	7.321	5.781	4.947	4.416	4.044	3.767	3.552	3.379	3.237	2.783	2.534
100K	10.828	6.908	5.422	4.617	4.103	3.743	3.475	3.266	3.098	2.959	2.513	2.266

K (multiply this value by 1,000)

Appendix 2-1: Four-Factor Screening Design

DESCRIPTION

This is a 12-run irregular* fraction (3/4 replicate) for four factors. The design allows you to estimate all main effects and two-factor interactions aliased only by three-factor or higher-order interactions. However, if effects are calculated hierarchically starting with main effects, these will be partially aliased with one or more interactions. In this case, be sure to review the probabilities in the ANOVA for the 2fi model. Exclude any main effects that are not significant.

*The number of runs is not a power of 2 (4, 8, or 16), as in a standard two-level factorial for four factors.

DESIGN LAYOUT

Std	A	B	C	D
1	−	−	−	−
2	+	+	−	−
3	−	−	+	−
4	+	−	+	−
5	−	+	+	−
6	+	+	+	−
7	−	−	−	+
8	+	−	−	+
9	−	+	−	+
10	+	+	−	+
11	+	−	+	+
12	−	+	+	+

ALIAS STRUCTURE FOR FACTORIAL MODEL

[Intercept]	=	Intercept − ABD
[A]	=	A − ACD
[B]	=	B − BCD
[C]	=	C − ABCD
[D]	=	D − ABCD
[AB]	=	AB − ABCD
[AC]	=	AC − BCD
[AD]	=	AD − BCD
[BC]	=	BC − ACD
[BD]	=	BD − ACD
[CD]	=	CD − ABD
[ABC]	=	ABC − ABD

ALIAS STRUCTURE FOR FACTORIAL MAIN-EFFECT MODEL

[Intercept]	=	Intercept − 0.333 * CD − 0.333 * ABC − 0.333 * ABD
[A]	=	A − 0.333 * BC − 0.333 * BD − 0.333 * ACD
[B]	=	B − 0.333 * AC − 0.333 * AD − 0.333 * BCD
[C]	=	C − 0.5 * AB
[D]	=	D − 0.5 * AB

Appendix 2-2: Five-Factor Screening Design

DESCRIPTION

This is a 16-run standard fraction (1/2 replicate) for five factors. The design allows you to estimate all main effects and two-factor interactions aliased only by three-factor or higher-order interactions.

DESIGN LAYOUT

Std	A	B	C	D	E
1	−	−	−	−	+
2	+	−	−	−	−
3	−	+	−	−	−
4	+	+	−	−	+
5	−	−	+	−	−
6	+	−	+	−	+
7	−	+	+	−	+
8	+	+	+	−	−
9	−	−	−	+	−
10	+	−	−	+	+
11	−	+	−	+	+
12	+	+	−	+	−
13	−	−	+	+	+
14	+	−	+	+	−
15	−	+	+	+	−
16	+	+	+	+	+

ALIAS STRUCTURE FOR FACTORIAL TWO-FACTOR INTERACTION MODEL

[Intercept]	=	Intercept + ABCDE
[A]	=	A + ABCD
[B]	=	B + BCDE
[C]	=	C + ABDE
[D]	=	D + ABCE
[E]	=	E + ABCD
[AB]	=	AB + CDE
[AC]	=	AC + BDE
[AD]	=	AD + BCE
[AE]	=	AE + BCD
[BC]	=	BC + ADE
[BD]	=	BD + ACE
[BE]	=	BE + ACD
[CD]	=	CD + ABE
[CE]	=	CE + BCD
[DE]	=	DE + ABC

Appendix 2-3: Six-Factor Screening Design

DESCRIPTION

This is a 32-run standard fraction (1/2 replicate) for six factors. The design allows you to estimate all main effects and two-factor interactions aliased only by four-factor or higher-order interactions. If you are willing to accept some problem-aliasing of two-factor interactions, consider adding one more factor to do seven in 32 runs.

DESIGN LAYOUT

Std	A	B	C	D	E	F
1	−	−	−	−	−	−
2	+	−	−	−	−	+
3	−	+	−	−	−	+
4	+	+	−	−	−	−
5	−	−	+	−	−	+
6	+	−	+	−	−	−
7	−	+	+	−	−	−
8	+	+	+	−	−	+
9	−	−	−	+	−	+
10	+	−	−	+	−	−
11	−	+	−	+	−	−
12	+	+	−	+	−	+
13	−	−	+	+	−	−
14	+	−	+	+	−	+
15	−	+	+	+	−	+
16	+	+	+	+	−	−
17	−	−	−	−	+	+
18	+	−	−	−	+	−
19	−	+	−	−	+	−
20	+	+	−	−	+	+

(continued)

DESIGN LAYOUT (Continued)

Std	A	B	C	D	E	F
21	−	−	+	−	+	−
22	+	−	+	−	+	+
23	−	+	+	−	+	+
24	+	+	+	−	+	−
25	−	−	−	+	+	−
26	+	−	−	+	+	+
27	−	+	−	+	+	+
28	+	+	−	+	+	−
29	−	−	+	+	+	+
30	+	−	+	+	+	−
31	−	+	+	+	+	−
32	+	+	+	+	+	+

ALIAS STRUCTURE FOR FACTORIAL TWO-FACTOR INTERACTION MODEL

All main effects and two-factor interactions are aliased only with four-factor or higher-order interactions, which aren't worth noting. This design also estimates 10 of the 20 three-factor interactions that are aliased with each other.

APPENDICES

Appendix 2-4: Seven-Factor Screening Design

DESCRIPTION:

This is a 32-run standard fraction (1/4 replicate) for seven factors. Main effects are clear of two-factor interactions. Most of the two-factor interactions are aliased only with 3-factor or higher interactions. However, the two-factor interactions involving D, E, F and G present problems with aliasing (see table below). Therefore, you would be wise to assign factors that are least likely to interact with the labels D, E, F or G, and those factors most likely to interact with A, B and C.

DESIGN LAYOUT

Std	A	B	C	D	E	F	G
1	−	−	−	−	−	+	+
2	+	−	−	−	−	−	−
3	−	+	−	−	−	−	−
4	+	+	−	−	−	+	+
5	−	−	+	−	−	−	−
6	+	−	+	−	−	+	+
7	−	+	+	−	−	+	+
8	+	+	+	−	−	−	−
9	−	−	−	+	−	−	+
10	+	−	−	+	−	+	−
11	−	+	−	+	−	+	−
12	+	+	−	+	−	−	+
13	−	−	+	+	−	+	−
14	+	−	+	+	−	−	+
15	−	+	+	+	−	−	+
16	+	+	+	+	−	+	−
17	−	−	−	−	+	+	−
18	+	−	−	−	+	−	+
19	−	+	−	−	+	−	+
20	+	+	−	−	+	+	−

(continued)

DESIGN LAYOUT *(Continued)*

Std	A	B	C	D	E	F	G
21	−	−	+	−	+	−	+
22	+	−	+	−	+	+	−
23	−	+	+	−	+	+	−
24	+	+	+	−	+	−	+
25	−	−	−	+	+	−	−
26	+	−	−	+	+	+	+
27	−	+	−	+	+	+	+
28	+	+	−	+	+	−	−
29	−	−	+	+	+	+	+
30	+	−	+	+	+	−	−
31	−	+	+	+	+	−	−
32	+	+	+	+	+	+	+

ALIAS STRUCTURE FOR FACTORIAL MODEL

[Intercept] = Intercept + DEFG + ABCDF + ABCEG
[A] = A + BCDF + BCEG + ADEFG
[B] = B + ACDF + ACEG + BDEFG
[C] = C + ABDF + ABEG + CDEFG
[D] = D + EFG + ABCF + ABCDEG
[E] = E + DFG + ABCG + ABCDEF
[F] = F + DEG + ABCD + ABCEFG
[G] = G + DEF + ABCE + ABCDFG
[AB] = AB + CDF + CEG + ABDEFG
[AC] = AC + BDF + BEG + ACDEFG
[AD] = AD + BCF + AEFG + BCDEG
[AE] = AE + BCG + ADFG + BCDEF
[AF] = AF + BCD + ADEG + BCEFG
[AG] = AG + BCE + ADEF + BCDFG
[BC] = BC + ADF + AEG + BCDEFG
[BD] = BD + ACF + BEFG + ACDEG
[BE] = BE + ACG + BDFG + ACDEF

[BF]	=	BF + ACD + BDEG + ACEFG
[BG]	=	BG + ACE + BDEF + ACDFG
[CD]	=	CD + ABF + CEFG + ABDEG
[CE]	=	CE + ABG + CDFG + ABDEF
[CF]	=	CF + ABD + CDEG + ABEFG
[CG]	=	CG + ABE + CDEF + ABDFG
** [DE]	=	DE + FG + ABCDG + ABCEF
** [DF]	=	DF + EG + ABC + ABCDEFG
** [DG]	=	DG + EF + ABCDE + ABCFG
[ADE]	=	ADE + AFG + BCDG + BCEF
[ADG]	=	ADG + AEF + BCDE + BCFG
[BDE]	=	BDE + BFG + ACDG + ACEF
[BDG]	=	BDG + BEF + ACDE + ACFG
[CDE]	=	CDE + CFG + ABDG + ABEF
[CDG]	=	CDG + CEF + ABDE + ABFG

** *Most troublesome aliases*

GLOSSARY

Glossary of Statistical Symbols

df   degrees of freedom
k   number of factors in design
i   individual datum
n   number of observations in sample
p   fraction of design (example 2^{k-p}) *or* probability value (ex. prob>F)
PI   prediction interval
r   sample correlation coefficient
R²   index of determination
s   sample standard deviation
s²   sample variance
t   t-value
X   independent variable
Y   observed response value
Z   uncontrolled variable
*   multiplication symbol
α   Type I risk ("alpha")
β   coefficient ("beta") or type II error.
—   average (bar)
Σ   mathematical operator to take the sum of a number series

Glossary of Terms

Actual value
1. The value of the response from the experiment.
2. The physical levels of the variables in the appropriate units, as opposed to their coded levels.

Adjusted R-squared
R-squared adjusted for the number of terms in the model relative to the number of points in the design. An estimate of the fraction of overall variation in the data accounted for by the model.

Alias
An effect (or model term) that is correlated with another effect. The resulting predictive model is then said to be "aliased."

Alpha (α)
See **Risk**.

Analysis of variance (ANOVA)
A statistical method, based on the F-test, that assesses the significance of experimental results. It involves subdividing the total variation of a set of data into component parts.

Antagonism
An undesirable interaction of two factors wherein the combination produces a response that is less than what would be expected from either one alone. The same concept can be applied to higher-order interactions.

Average
See **Mean**.

Axial points
Design points that fall on the spatial coordinate axes emanating from the overall centerpoint (or centroid in mixture space), often used as a label for star points in a central composite design.

GLOSSARY

Balanced design
Designs in which low and high levels of any factor or interaction occur in equal numbers.

Bias
A systematic error in estimation of a population value.

Block
A group of trials based on a common factor. Blocking is advantageous when there is a known factor that may influence the experimental result, but the effect is not of interest. For example, if all experiments cannot be conducted in one day or within one batch of raw material, the experimental points can be divided in such a way that the blocked effect is eliminated before computation of the model. Removal of the block effect reduces the noise in the experiment and improves the sensitivity to effects.

Case statistics
Diagnostic statistics calculated for each case; that is, each design point in the design after the model has been selected.

Categorical variable
Factors whose levels fall into discrete classes, such as metal versus plastic material.

Cell
The blank field to be filled with a response resulting from a given set of input factor levels.

Centerpoint
An experiment with all numerical factor levels set at their midpoint value.

Central composite design (CCD)
A design for response surface methods (RSM) that is composed of a core two-level factorial plus axial points and center points.

Central limit theorem
In its simplest form, this theorem states that the distribution of averages becomes normal as the sample size (n) increases. Furthermore, the variance of the averages is reduced by a factor of n when compared with the variance of individual data.

Centroid
The center point of mixture space within the specified constraints.

Class variable
See **Categorical variable**.

Coded factor level
See **Coding**.

Coding
A way to simplify calculations; in the case of two-level factorials, by converting low and high factor levels to –1 and +1, respectively.

Coefficient
See **Model coefficient**.

Coefficient of Variation (C V)
The coefficient of variation is a measure of residual variation of the data relative to the size of the mean. It is the standard deviation (root mean square error from ANOVA) divided by the dependent mean, expressed as a percent.

Component
An ingredient of a mixture.

Confidence interval
A data-based interval constructed by a method that covers the true population value a stated percentage (typically 95%) of the time in repeated samplings.

Confounding
See **Alias**.

Constrained
Limited in respect to component ranges for a mixture.

Continuous variable
See **Numerical variable**.

Contour plot
A topographical map drawn from a polynomial model, usually in conjunction with response surface methods for experimental design.

Cook's distance
A measure of how much the estimated coefficients would change if a particular run were omitted from the analysis. The values are relative. If a value is much higher than the others, it indicates that the case is influential. In other words, points with high Cook's distance carry a lot of weight in the fitting of the predictive model.

Corrected total
The total sum of squares (SS) corrected for the mean.

Count data
Data based on discrete occurrences rather than from a continuous scale.

Crash and burn
Exceed the operating boundaries (envelope) of a process.

Cumulative probability
The proportion of individuals in a population that the fall below a specified value.

Curvature
A measure of the offset at the centerpoint of actual versus predicted values from a factorial model. If it is significant, a quadratic model may be fitted to the data from a response surface design.

Degree of equation
The highest order of terms in a model. For example, in an equation of degree two, you will find terms with two factors multiplied together, as well as squared terms.

GLOSSARY

Degrees of freedom
The number of independent comparisons available to estimate a parameter.

Dependent mean
The mean of the response over all the design points.

Design matrix
An array of values presented in rows and columns. The columns usually represent design factors. The values in the rows represent settings for each factor in the individual runs or experiments of the design.

Design parameters
The number of treatments, factors, replicates, and blocks within the design.

Design space
An imaginary area bounded by the extremes of the tested factors.

Deterministic
An outcome that does not vary for a given set of input factors.

Distribution
A dispersion of data collected from a population.

Dot plot
A method for recording a response by simply putting points on a number line.

Effect
The change in average response when a factor, or interaction of factors, goes from its low level to its high level.

Envelope
The operating boundaries of a process.

Error term
The term in the model which represents random error. The data residuals are used to estimate the nature of the error term. The usual assumption is that the error term is normally and randomly distributed about zero, with a standard deviation of sigma.

GLOSSARY

Experiment
A test for the purpose of discovery.

Experimental region
See **Design space**.

F-test
See **F-value**.

F-Value
The F-distribution is a probability distribution used to compare variances by examining their ratio. If they are equal, the F-value is 1. The F-value in the ANOVA table is the ratio of model mean square (MS) to the appropriate error mean square. The larger their ratio, the larger the F-value and the more likely that the variance contributed by the model is significantly larger than random error.

F-Distribution
A probability distribution used in analysis of variance. The F-distribution is dependent on the degrees of freedom (df) for the variance in the numerator and the df of the variance in the denominator of the F-ratio.

Factor
The independent variable to be manipulated in an experiment.

Factorial design
A series of runs in which combinations of factor levels are included.

Fold-over
A method for augmenting low-resolution two-level factorial designs that requires adding runs with opposite signs to the existing block of factor(s).

Fractional factorial
An experimental design including only a subset of all possible combinations of factor levels, causing some of the effects to be aliased.

Full factorial
An experimental design including all possible combinations of factors at their designated levels.

GLOSSARY

General factorial
A type of full factorial that includes some categorical factors at more than two levels.

Half-normal
The normal distribution folded over to the right of the zero point by taking the absolute value of all data. Usually refers to a plot of effects developed by Daniel.

Hierarchy
The ancestral lineage of effects flowing from main effects (parents) down through successive generations of higher-order interactions (children). For statistical reasons, models containing subsets of all possible effects should preserve hierarchy. Although the response may be predicted without the main effects when using the coded variables, predictions will not be the same in the actual factor levels unless the main effects are included in the model. Without the main effects, the model will be scale-dependent.

Homogeneous
Uniform in composition, consisting of individuals which are of similar nature.

Hypothesis (H)
A mathematical proposition set forth as an explanation of a scientific phenomena.

Hypothesis test
See **Hypothesis**.

Identity column (I)
A column of all pluses in the design matrix, used to calculate the overall average. (Alternatively: intercept.)

Independence
A desirable statistical property where knowing the outcome of one event tells nothing about what will happen in another event.

Individuals
Discrete subjects or data from the population.

GLOSSARY

Interaction
The combined change in two factors that produces an effect greater (or less) than that of the sum of effects expected from either factor alone. Interactions occur when the effect one factor has depends on the level of another factor.

Intercept
The constant in the regression equation.

Irregular fraction
A two-level fractional factorial design that contains a total number of runs that is not a power of two. For example, a 12-run fraction of the 16-run full factorial design on four factors is a 3/4 irregular fraction.

Lack of fit (LOF)
A test that compares the deviation of actual points from the fitted surface, relative to pure error. If a model has a significant lack of fit, it is not a good predictor of the response and should not be used.

Lake Wobegon Effect
A phenomenon that causes all parents to believe their children are above the mean. It is named after the mythical town in Minnesota where, according to author Garrison Keillor, all the women are strong, all the men good looking, and all the children above average.

Least significant difference (LSD)
A numerical value used as a benchmark for comparing treatment means. When the LSD is exceeded, the means are considered to be significantly different. The LSD bars that appear on effect graphs are set at one-half the actual LSD. LSD bars that do not overlap indicate significant differences.

LSD bars
Plotted intervals around the means on effect graphs with lengths set at one-half the least significant difference. Bars that do not overlap indicate significant pair-wise differences between specific treatments.

Least squares
See **Regression analysis**.

GLOSSARY

Level
The setting of a factor.

Leverage
The potential for a design point to influence the fit model coefficients. Leverages near 1 should be avoided. If leverage is 1, then the model is forced to go through the point. Replicate such points to reduce their leverage.

Linear model
A polynomial model containing only linear or main-effect terms.

Lurking variable
An unobserved factor (one not in the design) causing a change in response. A classic example is the study of population in Oldenburg versus the number of storks, which led to the spurious conclusion that storks cause babies.

Main effect
The change in response caused by changing a single factor.

Mean
The sum of all data divided by the number of data—a measure of location.

Mean square
The sum of squares divided by the number of degrees of freedom (SS/DF). Analogous to a variance.

Median
The middle measurement.

Mixed factorial
See **General factorial**.

Mixture model
See **Scheffé polynomial**.

Mode
The value that occurs most frequently.

Model
An equation, typically a polynomial, that is fit to the data.

Model coefficient
The coefficient of a term in the regression model.

Multicollinearity
Multicollinearity problems arise when the predictor variables are highly interrelated; i.e., some predictors are nearly linear combinations of others. Highly collinear models tend to have unstable regression coefficient estimates.

Multiple response optimization
Method(s) for simultaneously finding the combination of factors giving the most desirable outcome for more than one response.

Normal distribution
A frequency distribution for variable data, represented by a bell-shaped curve symmetrical about the mean with a dispersion specified by its standard deviation.

Normal probability plot
A graph with a y-axis that is scaled by cumulative probability (Z), thus showing at a glance whether a given set of data is normally distributed.

Null
No difference.

Numerical variable
A quantitative factor that can be varied on a continuous scale, such as temperature.

OFAT
One factor at a time method of experimentation (as opposed to factorial design).

Observation
A record of factors and associated responses for a particular experimental trial.

Order

A measure of complexity of a polynomial model. For example, first-order models contain only linear terms. Second-order models contain linear terms plus two-factor interaction terms and/or squared terms. The higher the order, the more it can approximate curvature in response surfaces.

Orthogonal arrays

Test matrices exhibiting the property of orthogonality.

Orthogonality

A property of an experimental matrix that exhibits no correlation among its factors, thus allowing them to be estimated independently.

Outlier

A design point where the response does not fit the model.

Outlier t-test

Tests whether a run is consistent with other runs, assuming the chosen model holds. Model coefficients are calculated based on all design points except one. A prediction of the response at this point is made. The residual is evaluated using the t-test. A value greater than 3.5 means the point should be examined as a possible outlier.

p-value

Probability value, usually relating to the risk of falsely rejecting a given hypothesis.

Parameter

See **Model coefficient**.

Pencil test

A quick and dirty method for determining if a series of points fall on a line.

Plackett-Burman design

A class of saturated, orthogonal main effect, fractional two-level factorial designs where the number of runs is a multiple of four, rather than 2^k. These designs are Resolution III.

GLOSSARY

Poisson
A distribution characterizing discrete counts, such as the number of blemishes per unit area of a material surface.

Polynomials
Mathematical equations, comprised of terms with increasing order, used to approximate a true relationship.

Population
A finite or infinite collection of all possible individuals who share a defined characteristic; for example, all parts made by a specific process.

Power
The probability that a test will reveal an effect of stated size.

Power law
A relationship between the true standard deviation and the true mean to some power. Non-zero powers cause a violation of basic statistical assumptions that can be rectified via a response transformation. Ideally the power is zero, indicating there is no relationship, so no transformation is required.

Predicted R-squared
Measures the amount of variation in new data explained by the model. It makes use of the predicted residual sum of squares (PRESS) as shown in the following equation: Predicted R-squared = $1 - SS_{PRESS}/(SS_{TOTAL} - SS_{BLOCKS})$.

Predicted value
The value of the response predicted by the mathematical model.

Predicted residual sum of squares (PRESS)
A measure, the smaller the better, of how well the model fits each point in the design. The model is repeatedly refitted to all the design points except the one being predicted. The difference between the predicted value and actual value at each point is then squared and summed over all points to create the PRESS.

Prob >F (probability of a larger F-value)

If the F-ratio—the ratio of variances—lies near the tail of the F-distribution, the probability of a larger F is small, and the variance ratio is judged to be significant. Usually, a probability less than 0.05 is considered significant. The F-distribution is dependent on the degrees of freedom (df) for the variance in the numerator and the df of the variance in the denominator of the F-ratio.

Prob >t (probability of a larger t-value)

The probability of a larger t-value if the null hypothesis is true. Small values of this probability indicate significance and rejection of the null hypothesis.

Probability paper

Graph paper with specially scaled y-axis for cumulative probability. The purpose of the paper is to display normally distributed data as a straight line. It's used for diagnostic purposes to validate the statistical assumption of normality.

Process

Any unit operation, or series of unit operations, with measurable inputs and outputs (responses).

Pure error

Experimental error, or pure error, is the normal variation in the response, which appears when an experiment is repeated. Repeated experiments rarely produce exactly the same results. Pure error is the minimum variation expected in a series of experiments. It can be estimated by replicating points in the design. The more replicated points, the better will be the estimate of the pure error.

Quadratic

A second-order polynomial.

Qualitative

See **Categorical**.

Quantitative

See **Numerical**.

GLOSSARY

R-squared

The multiple correlation coefficient which provides an estimate of the fraction of overall variation in the data accounted for by the model. It is a number between zero and one that indicates the degree of relationship of the response variable to the combined linear predictor variables.

Randomization

Mixing up planned events so they follow no particular pattern—important in ensuring that lurking variables do not bias the outcome. Randomization of the order in which experiments are run is essential to satisfy the statistical requirement of independence of observations.

Range

The difference between the largest and smallest value—a measure of dispersion.

Regression analysis

A method by which data is fitted to a mathematical model.

Replicate

An experimental run performed again from start to finish (not just resampled and/or remeasured.) Replication provides an estimate of pure error in the design.

Residual

See **Residual error**.

Residual error

The difference between the observed response and the value predicted by the model for a particular design point.

Response

A measurable product or process characteristic thought to be affected by experimental factors.

Response surface method (RSM)

A statistical technique for modeling responses via polynomial equations. The model becomes the basis for 2-D contour maps and 3-D surface plots for purposes of optimization.

GLOSSARY

Risk
The probability of making an error in judgment, i.e., falsely rejecting the null hypothesis. (See also **Significance level**.)

Root mean square error
The square root of the residual mean square error. It estimates the standard deviation associated with experimental error.

Rule-of-thumb
A crude method for determining whether a group of points exhibit a nonrandom pattern: if, after covering any point(s) with your thumb(s), the pattern disappears, there is no pattern.

Run order
Run order is the randomized order for experiments. Run numbers should start at one and include as many numbers as there are experiments. Runs must be continuous within each block.

Sample
A subset of individuals from a population, usually selected for the purpose of drawing conclusions about specific properties of the entire population.

Saturated
An experimental design with the maximum number of factors allowable without aliasing their main effect.

Scheffé polynomial
A form of mathematical predictive model designed specifically for mixtures. They are derived from standard polynomials, of varying degrees, by accounting for the mixture constraint that all components sum to the whole.

Screening
Sifting through a number of variables to find the vital few. Two-level fractional factorial designs are often chosen for this purpose.

Significance level
The level of probability, usually 0.05, established for rejection of the null hypothesis.

GLOSSARY

Simplex

A geometric figure with one more vertex than the number of dimensions. For example, an equilateral triangle.

Simplex centroid

A mixture design comprised of the purest blends, binary combinations, etc., up to and including a centroid blend of all components.

Sparsity of effects

A rule-of-thumb that about 20% of main effects and two-factor interactions will be active in any given system. The remainder of main effects, two-factor interactions, and all higher-order effects are near zero, with a variation based on underlying error.

Standard deviation

A measure of variation in the original units of measure, computed by taking the square root of the variance.

Standard error of a parameter

The estimated standard deviation of a parameter or coefficient estimate.

Standard error

The standard deviation usually associated with the mean rather than individuals.

Standard order

A conventional ordering of the array of low and high factor levels versus runs in a two-level factorial design.

Star points

Extreme axial points in a central composite design.

Statistic

A quantity calculated from a sample to make an estimate of a population parameter.

Studentized

A value divided by its associated error. The resulting quantity is a dimensionless score useful for purposes of comparison.

Studentized residual
The residual divided by the estimated standard deviation of that residual.

Stuff
Processed material, such as food, pharmaceuticals, or chemicals (as opposed to **Thing**).

Sum of squares (SS)
The sum of the squared distances from the mean due to an effect.

Synergism
A desirable interaction of two factors where the combination produces a response that is greater than what would be expected from either one alone. The same concept may be applied to higher-order interactions.

Thing
Manufactured hard goods, such as electronics, cars, or medical devices (as opposed to **Stuff**).

t-value
A value associated with the t-distribution that measures the number of standard deviations separating the parameter estimate from zero.

Tetrahedron
A three-dimensional geometric figure with four vertices. It is a simplex. The tetrahedron looks like a pyramid, but has only three sides, not four.

Transformation
A mathematical conversion of response values.

Treatment
A generic term borrowed from agricultural research that originally meant two or more fertilizers or other factors being compared in a field trial.

Trivial many
Nonactive effects. The sparsity of effects principal predicts that all interactions of third-order or higher will fall into this category, as well as 80 percent of all main effects and two-factor interactions.

GLOSSARY

True
Related to the population rather than just the sample.

Trial
See **Experiment**.

Type 1 error
Saying something happened when it really didn't (a false alarm).

Type 2 error
Not discovering that something really happened (failure to alarm).

Uniform distribution
A frequency distribution where the expected value is a constant, exhibiting a rectangular shape.

Variable
A quantity that may assume any given value or set of values.

Variance
A measure of variability computed by summing the squared deviations of individual data from their mean, then dividing this quantity by the degrees of freedom.

Vertex
A point representing an extreme combination of input variables subject to constraints. Normally used in conjunction with mixture components.

Vital few
The active effects. (Also see **Sparsity of effects**.)

X-space
See **Design space**.

Y-bar
A response mean.

RECOMMENDED READINGS

Textbooks

1. Amsden, Robert, and Howard Butler. *SPC Simplified, Practical Steps to Quality*. New York: Quality Resources, 1986.
2. Box, George, J. Stuart Hunter, and William Hunter. *Statistics for Experimenters*. New York: John Wiley and Sons, 1978.
3. Cornell, John. *Experiments with Mixtures*, 2nd ed. New York: John Wiley and Sons, 1990.
4. Gonick, Larry, and Woollcott Smith. *The Cartoon Guide to Statistics*. New York: HarperCollins, 1993.
5. John, Peter. *Statistical Design and Analysis of Experiments*. New York: The Macmillan Company, 1969.
6. Montgomery, Douglas. *Design and Analysis of Experiments*, 4th ed. New York: John Wiley and Sons, 1997.
7. Montgomery, Douglas, and Raymond Myers. *Response Surface Methodology*, New York: John Wiley and Sons, 1995.
8. Phillips, John. *How to Think About Statistics*, 2nd edition. New York: W. H. Freeman, 1982.

Case-Study Articles

Reprints for most of the articles cited below can be obtained by e-mail request to **info@statease.com**. Please specify which reprint(s) are desired. Include postal address in the e-mail. Also check our web site at **www.statease.com** for newer reprints (on-line or downloadable.)

1. Allnoch, Allen. "DOE Package Optimizes Coverwrap Process," *Industrial Engineering Solutions* (June 1997): 56–57. DOE minimizes scrap, rework and cleaning times caused by excessive glue buildup during photo album coverwrap process.
2. Anderson, Mark. "Design of Experiments," *The Industrial Physicist* (September 1997): 24–26. General review of factorial design with several brief case studies and a listing of DOE software vendors.

RECOMMENDED READINGS

3. Anderson, Mark and Paul Anderson. "Design of Experiments for Process Validation," *Medical Device and Diagnostic Industry* (January 1999): 193–199. A detailed case study on how to use low-resolution two-level factorials, called "ruggedness tests," for validation of medical devices or other finished products, methods, or systems.
4. Anderson, Mark and Patrick Whitcomb. "Breakthrough Improvements with Experiment Design," *Rubber and Plastics News* (June 16, 1997): 14–15. Detailed case study on a fractional two-level factorial for injection molding.
5. Anderson, Mark and Shari Kraber. "Eight Keys to Successful DOE," *Quality Digest* (July 1999): 39–43. A general article on the basic elements of factorial design.
6. Anderson, Mark and Patrick Whitcomb. "Computer-Aided Design of Experiments for Formulations," *Modern Paint and Coatings* (June 1997): 38–39. Mixture design for additive package for viscosity modification.
7. Anderson, Mark and Patrick Whitcomb. "Design of Experiments Strategies," *Chemical Processing* (1998 Project Engineering Annual): 46–48. Detailed how-to article on application of two-level factorials and response surface methods to a catalytic chemical process.
8. Anderson, Mark and Patrick Whitcomb. "Find the Most Favorable Formulations," *Chemical Engineering Progress* (April 1998): 63–67. Detailed how-to article on applying mixture design to a chemical formulation.
9. Anderson, Mark and Patrick Whitcomb. "Mixing It Up with Computer-Aided Design," *Today's Chemist at Work*, Vol. 6, No. 10 (November 1997): 34–38. Detailed how-to article on applying mixture design to food formulation.
10. Anderson, Mark and Patrick Whitcomb. "Optimization of Paint Formulations Made Easy with Computer-Aided Design of Experiments for Mixtures," *Journal of Coatings Technology* (July 1996): 71–75. Essential aspects of mixture methodology are illustrated in a case study on latex paint.
11. Anderson, Mark and Patrick Whitcomb. "Optimize Your Process-Optimization Efforts," *Chemical Engineering Progress* (December 1996): 51–60. Detailed how-to article on applying two-level factorials and response surface methods to a chemical process.
12. Anderson, Mark. "Prototype Circuit Board Maker DOEs Gold Plating," *Job Shop Technology* (July 1998): 66–70. A circuit board maker runs two different factorial designed experiments to reduce rework and improve yields.
13. Anderson, Mark and Patrick Whitcomb. "Software Sleuth Solves Engineering Problems," *Machine Design* (June 5, 1997): 62–66. Detailed article showing the power of two-level factorials to reveal interactions in the design of a bearing. End result is a breakthrough increase in part life.

RECOMMENDED READINGS

14. Anderson, Mark, P.K. Bhattacharjee, and Patrick Whitcomb, "Statistical Design of Experiments for Quality Improvement of Fertilizer Products," *American Institute of Chemical Engineering Proceedings* (Spring 1998). Factorial design maximizes nitrogen content in fertilizer, making it less susceptible to impurities in lower-grade phosphates.
15. Cottington, L. James. "How Wacker Uses DOE Software to Optimize Silicone Elastomer Formulations," *Today's Chemist at Work*, Vol. 5, No. 11, (December 1996): 39–40. Mixture design is applied and multiple responses optimized.
16. Blattenberger, David, and Mary Bertelson. "Development of a Multicomponent Allergy Screen Assay," *Biotechnology International*, (1999): 123–127. Response surface methods are applied and multiple responses optimized.
17. Chase, Nancy. "Experiments Uncover Source of Valve Failures," *Quality* (June 1998): 69–72. A manufacturing company reduces overall product development time by testing design, environmental, and manufacturing factors simultaneously.
18. De Vowe, David. "Diecaster Achieves Zero-Defect Parts," *Quality in Manufacturing* (April 1994): 20. DOE reduces incidence of defects in diecast aluminum housings for computer disk drives from an average 15 percent to zero percent.
19. Editor, "Design of Experiments Reveals Quality Breakthroughs," *The Quality Observer* (January 1996): 7–10. Supported by brief case studies, article shows why DOE is an indispensable tool for quality engineers.
20. Editor, "Designed Experiments Aid Specialty Manufacturer," *Scientific Computing & Automation* (September 1996): 13–14. DOE reduces time to develop a variety of products.
21. Editor, "Solving Core Shear with Design of Experiments," *Job Shop Technology* (March 1999): 37–41. A foundry did a series of three, two-level factorial designs to reduce rework on cast bolt-holes. They save $50,000 per year.
22. Giacobbe, Thomas and Thomas O'Brien. "Faster, More Efficient Cutting Fluid Formulation," *MAN-Modern Application News* (July 1995): 40–41. DOE enables experimenters to test 28 variables in four weeks and develop a new metalworking fluid that gives better machinability.
23. Hazelwood, Steve. "Using DOE to Prevent Solvent Pop," *Paint & Coatings Industry* (August 1998): 96–100. DOE reveals optimum factors to increase paint film thickness without causing solvent pop or blistering.
24. Hubbell, Douglas. "DOE Software Helps K2 Design Cap Skis," *Quality* (November 1995): 46–48. DOE reduces scrap due to blisters, dimples, and delamination of fiberglass skis.

RECOMMENDED READINGS

25. Keller, Maureen. "Perfecting the Ice Cube," *Scientific Computing & Automation* (May 1995): W30–W32. DOE provides in-depth details on three factors affecting production of ice in refrigerators.

26. Kim, John and James Kalb. "Design of Experiments: An Overview and Application Example." *Medical Device & Diagnostic Industry* (March 1996): 78–88. Detailed overview shows how to apply DOE to medical product design, development, and validation.

27. Koselka, Rita. "The New Mantra: MVT," *Forbes* (March 11, 1996): 114–118. How to apply multivariable testing to nonmanufacturing areas such as mail-order or hospitals.

28. Krahnke, Bob, "Designed Experimentation Reveals Best Approach for Silicone Sealants," *Adhesives Age* (August 1994): 30–32. Researchers use response surface methods to improve quality and reduce costly ingredients to create high-quality products.

29. Mayes, Richard and John Ganjei. "Statistical Software Speeds Process Troubleshooting," *Chemical Engineering* (December 1995): 115. DOE reduces photoresist adhesion problems during printed circuit board manufacturing.

30. Mills, Wayne. "John Deere Saves $500K Annually with DOE," *Scitech Journal* (July 1995): 18–19. DOE reveals that expensive additive of chromate conversion is not a significant factor in paint adhesion. Instead, paint type is identified as a far more important factor for improving paint adhesion.

31. Remily, Nicole. "Software Optimizes Nabisco Cookie Line," *Food Engineering Magazine* (May 1997): 51–54. DOE helps eliminate scrap and improve product shape.

32. Runke, Peter. "Design of Experiments Software Saves Kodak Thousands," *Metal Forming Magazine* (April 1996): 37–40. DOE reveals that new tooling on an existing stamping machine holds a metal forming operation to high repeatability and tolerances.

33. Smith, Jr., Wendell. "Mixture Experiments Hold the Keys to Formulation," *Today's Chemist at Work*, Vol. 5, No. 2, (February 1996): 18–24. Detailed how-to article on applying mixture design to formulation.

34. Stohr, Doug. "Design of Experiments Finds Vital Factors in Complex Processes," *Metal Forming Magazine* (December 1997): 31–33. The dilemma to retool or to buy a new metalforming machine confirms DOE's systematic approach. Findings reveal that retooling holds the existing machine to high repeatability. Clear graphics and tables present a breakthrough that's easy to understand.

INDEX

Aliasing effects, 87–88, 90–91, 92–97
  blocking and, 96, 115
  confetti example, 140
  dancing raisin example, 112–14, 115–19
  fold-over design augmentation and, 115–19
  irregular fractions and, 101
  minimal-run designs and, 109, 119–20
  molding example, 106, 171
  in Plackett-Burman two-level designs, 98, 115
  in screening designs, 191, 193, 195, 197–98
  skating example, 173, 175–76
ANOVA (analysis of variance), 24
  car temperature example, 162–63
  commuting example, 177
  computer screen readability example, 102
  confetti example, 139, 142
  gold-copper amalgam example, 149
  in Microsoft Excel 2000, 10
  microwave popcorn example, 58–61
  Slinky spring examples, 126, 130
  tabletop hockey example, 79
Antagonism, 148
  response surfaces and, 153
Anti-effect, 114
Antilog, 80, 83
Archimedes, 150
Arrhenius, S.A., 83
Averages. *See* Means

Ball in funnel experiment, 183
Balsa airplane experiment, 183
Battery life example, 132–33, 176
Bell-shaped distributions. *See* Normal distributions
Bias, 29
Blocking, 32
  dice sorting example, 32–36
  fold-over design augmentation and, 115
  in fractional factorials, 96
  F-test and, 34–35
  molding example, 106, 170
  randomized, 32
  by repeated trials, 33
  shoe wear example, 32
Bowling team example, 37, 157
Box, George, 94

Car temperature example, 70–71, 162–64
CCD (central composite design), 140–44
Centerpoints, 136–38
Central composite design. *See* CCD (central composite design)
Central limit theorem, 7
  F-test and, 23
  normal distributions, 7
  standard error and, 11
Chance patterns, 25, 67
Commuting example, 133–34, 176–78
Computer screen readability example, 99–105
Confetti example, 136–40
  with RSM, 140–44
Confidence intervals, 12–16. *See also* LSD (least significant difference)
  dice example, 15
  error probability and, 16
  p-values, 24
  t-statistic and, 12–15
Confounding relationships. *See* Aliasing effects
Contour graphs, 143
  teeny beany example, 152
Copper-gold blend example. *See* Gold-copper amalgam example

INDEX

Cube plots
  computer screen readability example, 105
  dancing raisin example, 113
  direct marketing example, 169–70
  microwave popcorn example, 47
  skating example, 174
  tabletop hockey example, 83
Cumulative probability, 16–17
  standard deviation vs., 17
Curvature, 139, 140

Dancing raisin example, 110–14
  aliasing effects, 112–14
  with fold-over design augmentation, 115–19
Degrees of freedom, 10
  dice examples, 29–30
  error and, 101
  interdependence of factors and, 149
  normal distribution and, 30–31
Design and Analysis of Experiments, 5, 68, 132
Design augmentation
  CCD, 140–44
  fold-over design augmentation, 98, 109, 115–19, 121–22, 175–76
  RSM, 135–44
Design of experiments. *See* DOE (design of experiments)
Dice example, 7–11, 15
  with color, 24–26, 38, 158
  sorting, 32–36
  with tampering, 28–32
Direct marketing example, 84–86, 165–70
  square root transformation, 165–69
DOE (design of experiments)
  Fisher and, 23
  flowchart, xiii
  level setting and, 44
  SPC (statistical process control) vs., 4
  systematic improvement and, 4

Effects plots. *See* One-factor plots

Errors
  degrees of freedom and, 101
  lack of fit, 139, 140
  percent, 77
  process vs. measurement, 137
  residual, 53, 58–59
  standard, 11–14
  Type I, 2, 28, 112
  Type II, 2
  Type III, 5
Examples
  battery life, 132–33, 176
  bowling team, 37, 157
  car temperature, 70–71, 162–64
  commuting, 133–34, 176–78
  computer screen readability, 99–105
  confetti, 136–44
  dancing raisin, 110–19
  dice, 7–11, 15
  dice sorting, 32–36
  dice with color, 24–26, 38, 158
  dice with tampering, 28–32
  direct marketing, 84–86, 165–70
  fabric wear, 39–40, 160–61
  gold-copper amalgam, 146–49
  gold-silver alloy, 150
  microwave popcorn, 43–68, 92–94
  motor-shaft endplay, 20, 38–39, 155–56, 159–60
  shoe wear, 32
  skating, 120–22, 173–76
  Slinky spring, 124–32
  suppliers, 20, 38–39, 155–56, 159–60
  teeny beany, 150–53
  weed whacker, 88–92
  weight, 18–20, 21, 156–57
Experiments
  ball in funnel, 183
  flight of the balsa buzzard, 183
  hand-eye coordination, 181–82
  impact craters, 184
  paper airplanes, 183
  paper clips, 179–80

INDEX

Fabric wear example, 39–40, 160–61
Factorial approach, 74
Factorial designs, 135
    centerpoints, 136–38
    flowchart, xiii
    fractional, 87–105, 110, 170–72, 190–98
    full, 87, 135
    general, 123–34
    hierarchy, 102
    interactions, 53, 54–57, 88–89, 102, 128
    interactions and residuals, 128, 130
    irregular fractions, 98–105
    minimal-run designs, 109–22, 173–76
    mixture, 145–53
    multilevel screening designs, 190–98
    one factor, xii, 23–40, 157–61
    Plackett-Burman two-level designs, 98, 115
    resolution, 92–97
    three-level, 138
    two-level, xii, 41–71, 136, 162–64
False negative errors, 2
False positive errors, 2, 28, 112
Family unity, 102, 113, 116
F-distributions, 160
Fisher, Sir Ronald, 23, 32
Flight of the balsa buzzard experiment, 183
Flowchart, xiii
Fold-over design augmentation
    aliasing effects, 115–19
    blocking, 115
    complete, 109, 115–17, 119
    dancing raisin example, 115–19
    resolution and, 98, 115–19
    single factor, 118–19
    skating example, 121–22, 175–76
Fractional factorials, 87, 135. *See also* Minimal-run designs
    aliasing effects, 87–88, 90–91, 92–97, 101
    blocking, 96
    computer screen readability example, 99–105
    in design for use, 110
    hierarchy, 102, 113, 116

    interactions, 88–89, 102, 128
    interactions and residuals, 128, 130
    irregular, 98–105
    microwave popcorn example, 92–94
    minimal-run designs, 109–22, 173–76
    molding example, 105–7, 170–73
    projection, 92
    resolution, 92–97
    screening designs, 190–98
    weed whacker example, 88–92
F-ratios, 24, 59
    values, 186–89
F-tests, 23–31. *See also* ANOVA
    blocking and, 34–35
    bowling team example, 37, 157
    central limit theorem and, 23
    dice example with color, 24–26, 38, 158
    dice example with tampering, 28–30
    dice sorting example, 34–35
    fabric wear example, 39–40, 160–61
    LSD and, 28
    microwave popcorn example, 59–60
    suppliers example, 38–39, 159–60
    variance, 24

General factorial designs, 123–34
    battery life example, 132–33, 176
    commuting example, 133–34, 176–78
    RSM and, 131
    Slinky spring examples, 124–32
    Gold-copper amalgam example, 146–49
Gold-silver alloy example, 150
Gosset, W.S., 14
Graphs. *See specific types of graphs and plots*
Grass cutter example. *See* Weed whacker example

Half-normal paper, 49, 50
    Half-normal plots, 16, 49, 50. *See also* Normal plots
    car temperature example, 162
    computer screen readability example, 100
    confetti example, 138

INDEX

dancing raisin example, 112, 116
direct marketing example, 165, 166
microwave popcorn example, 49–51, 54
molding example, 171
skating example, 174, 175
tabletop hockey example, 77, 79
weed whacker example, 90
Hand-eye coordination experiment, 181–82
Hedonic scale, 56
Hierarchy, preservation of, 102, 113, 116
Hockey example. *See* Tabletop hockey example
Hoppe, Frank August, 146
Hunter, Bill, 5
Hypothesis testing, 26

Impact craters experiment, 184
Induction, 33
Injection molding example. *See* Molding example
In-line skates example. *See* Skating example
Interaction graphs
   car temperature example, 164
   commuting example, 177
   computer screen readability example, 103
   dancing raisin example, 114, 117–18
   direct marketing example, 169
   microwave popcorn example, 55
   molding example, 172–73
   skating example, 176
   Slinky spring examples, 127, 131
   tabletop hockey example, 82
Interactions of factors, 53, 54–57, 88–89, 102, 128
   residuals and, 128, 130
Inverse transformations of data, 82
Irregular fractions, 98–105
   aliasing effects and, 101

James, Richard, 125
Jelly bean example. *See* Teeny beany example
Juran, Joseph, 52

Lack of fit, 139, 140
Least significant difference. *See* LSD (least significant difference)
Linear transformations of data, 44
Logarithms
   as scaling functions, 78
   tabletop hockey example, 74, 76
Logarithm transformations of data, 73, 74, 83
LSD (least significant difference), 27–28. *See also* Interaction graphs
   dice example with color, 27–28, 38, 158
   dice example with tampering, 30–32
   dice sorting example, 36
   F-test and, 28
   suppliers example, 159

Means, 7–9
Mean square, 59
Mean square residuals, 161
Measurement error vs. process error, 137
Microsoft Excel 2000, variance calculation in, 10
Microsoft Windows, calculator and statistics, 9
Microwave popcorn example, 43–54, 61
   ANOVA, 58–61
   factor interaction, 53, 54–57
   fractional factorials, 92–94
   modeling, 58, 62–63
   residuals, 63–68
Minimal resolution designs. *See* Minimal-run designs
Minimal-run designs, 109–22, 173–76
   dancing raisin example, 110–19
   resolution and aliasing, 109, 119–20
   skating example, 120–22, 173–76
Mixed factorials. *See* General factorial designs
Mixture design, 145–53
   constrained total, 146
   gold-copper amalgam example, 146–49
   gold-silver alloy example, 150

INDEX

proportions and, 145
teeny beany example, 150–53
Modeling
  coded vs. uncoded, 63
  confetti example, 140–44
  logarithms in, 74, 83
  microwave popcorn example, 58, 62–63
  residuals, 63–68
Modes, 7
Molding example, 105–7, 170–73
Montgomery, Douglas, 68, 132
Motor-shaft endplay examples, 20, 38–39, 155–56, 159–60
Multivariable testing. *See* DOE (design of experiments)
MVT (multivariable testing). *See* DOE (design of experiments)

Non-normal distributions. *See* Uniform distributions
Nonorthogonality, 46
Normal distributions, 5–8, 156
  central limit theorem, 7
  degrees of freedom and, 30–31
  sample size and, 13–15
  skewed distributions vs., 21
  t-distributions vs., 13–15
  uncontrolled variables and, 7
Normal plots, 16–20. *See also* Half-normal plots
  car temperature example, 164
  direct marketing example, 166–67
  microwave popcorn example, 64–66
  "pencil test," 19, 65
  of residuals, 64–66, 77, 80, 164, 166–67
  tabletop hockey example, 77, 80
  weight examples, 18–20, 21, 157
Null hypothesis, 26

OFAT (one-factor-at-a-time), two-level factorial vs., 41–43, 54, 168
Olaf of Norway, King, 6

One-factor-at-a-time analyses. *See* OFAT (one-factor-at-a-time)
One-factor plots
  computer screen readability example, 104
  confetti example, 139
  dice example with color, 27, 158
  dice example with tampering, 31
  dice sorting example, 36
  direct marketing example, 168
  fabric wear example, 161
  Slinky spring example, 132
  suppliers example, 159
  tabletop hockey example, 81
Optimal design, 131–32
Optimization, RSM, 135–44
Orthogonal arrays, 46
Orthogonality, 48
Outliers, 6

Paper airplanes experiment, 183
Paper clips experiment, 179–80
Pareto, Vilfredo, 52
Patterns, chance, 25, 67
"Pencil test," 19, 65
Percent errors, 77
Plackett-Burman arrays, 46
Plackett-Burman two-level designs, 98, 115
Plots. *See* specific types of graphs and plots
Poisson distributions, 83
Postcard example. *See* Direct marketing example
Power law relationship, transformations of data, 81–82
Probability
  cumulative, 16–17
  p-value, 24
  standard deviation vs., 17
Probability paper, 16, 18
Process error vs. measurement error, 137
Projection, 92
p-values, 24
  null hypothesis, 26
  Type I error and, 112

INDEX

Randomized blocking, 32
Randomizing
  run order, 45, 147
  uncontrolled variables, 25, 32
Rating scales, 56
Reactor filtration example, 68–70, 162
Repeated trials
  blocking and, 33
  in Plackett-Burman designs, 98
  replication and, 137
Replication, 137
Residuals, 53
  chance patterns in, 67
  interactions of factors and, 128, 130
  microwave popcorn example, 58–59, 63–68
  normal plots, 64–66, 67, 77, 80, 164, 166–67
  vs. predicted response, 66–68, 78, 80, 166, 167
  transformations of data and, 83
Resolution, 92–97
  anti-effect and, 114
  fold-over design augmentation and, 98, 115–19
  minimal-run designs and, 109, 119–20
Response surface methods. *See* RSM (response surface methods)
Response surfaces, 135, 136
  confetti example, 144
  gold-copper amalgam example, 148
  lack of fit, 139, 140
  teeny beany example, 153
Reverse transformations, 80, 83, 84, 168
RGB readability example. *See* Computer screen readability example
Risk. *See* P-value
RSM (response surface methods), 135–44
  CCD, 140–44
  lack of fit, 139, 140
  quantifying variables, 131
  response surfaces, 135, 136
  two-level designs and, 136, 140–44

Ruggedness testing, 110
"Rule of thumb" for residual plots, 67
Run-chart, variability and, 4

Saddle point response surface, 135, 136
Sampling, 11
  extrapolation from, 85
  size and distribution, 13–15, 27
  size and significance, 85
  standard error and, 11–12
Saturated designs, 109
Scaling functions, 78
Scatter-plots
  dice stacking example, 34
  molding example, 172
Scheffé, Henri, 147
Screening designs for fractional factorials, 190–98
Shoe wear example, 32
Simple maximum response surface, 135, 136
Simplex, 151
Simplex centroid, 150
Single factor fold-over design augmentation, 118–19
Skating example, 120–21, 173–75
  with fold-over design augmentation, 121–22, 175–76
Skewed distribution, 21
Slinky spring examples
  four springs, 128–32
  quantifying variables, 131
  RSM, 131
  three springs, 124–27
Sparsity of effects, 52
SPC (statistical process control), xi
  DOE (design of experiments) vs., 4
  variability assessment, 4, 24
SPC Simplified, xi
Spring toy examples. *See* Slinky spring examples
Square plot, weed whacker example, 91
Square root transformations of data, 82, 165–66

INDEX

Standard arrays, 46
Standard deviation, 10–11
  cumulative probability vs., 17
  Poisson distributions, 83
  of populations, 11
  power law relationships and, 82
  sampling, 11
  t-statistic and, 12–15
Standard error, 11–12
  t-statistics and, 12–14
Statistical process control. *See* SPC (statistical process control)
Statistics. *See also* ANOVA
  bias, 29
  categorical inputs and, 3–4, 43
  central limit theorem, 7, 11
  chance, 25
  confidence interval, 12–16
  degrees of freedom, 10
  descriptive, 1, 7–11
  and determining risk, 1, 12–16
  dice example, 7–11, 15
  graphical tests, 16–20
  mean, 7–9
  mode, 7
  normal distribution, 5–8, 13–15
  normal distributions, 16–20
  null hypothesis, 26
  numerical inputs, 3–4, 43
  sampling, 11
  standard deviation, 10–11, 12–15
  standard error, 11–14
  as summary, 1, 7–11
  suppliers examples, 20, 38–39, 155–56, 159–60
  t-statistic, 12–14, 185
  uncontrolled variables and, 3, 4, 32
  variance, 9–10
  weight examples, 18–20, 21, 156–57
Student (W.S. Gosset), 14
Suppliers examples, 20, 38–39, 155–56, 159–60
Synergism, 147, 148
  response surfaces and, 153

Tabletop hockey example, 73–77, 78
  logarithm transformation, 74, 76, 77–81
  modeling, 74
Taguchi arrays, 46
t-distributions
  degrees of freedom and, 30–31
  normal distributions vs., 13–15
  sample size and, 13–15
Teeny beany example, 150–53
Theorem, central limit, 7, 11
Thinking, as trial-and-error process, 33
Time-related variables, 45
Transformations of data, 83–84
  inverse, 82
  linear, 44
  logarithm, 73, 74, 76, 77–81, 83
  power law relationship and, 81–82
  residual variance and, 83
  reverse, 80, 83, 84, 168
  square root, 82, 165–66
Triangular graphs, teeny beany example, 151, 152
Trilinear graph paper, 151
t-statistics, 12–14
  values, 185
Two-level factorial designs, xii, 41–43. *See also* Design augmentation; Fractional factorials
  CCD, 140–44
  centerpoints, 136–38
  confetti example, 136–44
  curvature in, 139, 140
  gold-copper amalgam example, 146–49
  gold-silver alloy example, 150
  interactions, 53, 54–57
  microwave popcorn example, 43–54
  mixtures, 146–53
  one factor vs., 41–43, 54, 168
  orthogonal arrays, 46
  RSM and, 136, 140–44
  screening designs, 190–98
  sparsity of effects, 52
  three-level vs., 138

INDEX

Uncontrolled variables, 3, 4
   normal distribution and, 7
   randomizing, 25, 32
   time-related, 45
Uniform distributions, 5, 25

Variance, 9–10
   calculation in Microsoft Excel 2000, 10
   corrected for mean, 10, 59, 60, 163
   F-test, 24
Variation blocking, 32
Von Bortkiewicz, 83

Weed whacker example, 88–92
Weight examples, 18–20, 21, 156–57

ABOUT THE SOFTWARE

To make it easy to perform DOE, we've provided a limited-time version of a commercially available software package from Stat-Ease, Inc. called Design-Ease®. This Windows-based program comes with companion files that provide tutorials on one-factor, factorial, and general factorial designs. You will also find files of data for most of the exercises in the book. The data sets are named so they may be easily cross-referenced with the book. For example "Table2-1-dice.de6" refers to the Table 1 in Chapter 2. You are encouraged to reproduce the results shown in the book, plus do further exploration. The Stat-Ease software offers far more detail in statistical outputs and many more graphics than can be included in this book. You will find a great deal of information on program aspects and statistical background in the on-line hypertext help system built into the software.

Refer to the enclosed instructions for a step-by-step procedure for installing the software, the tutorials, and data files. The tutorials come in PDF format (Portable Document Files), which can be read with Adobe's Acrobat Reader (included on the CD-ROM).

Technical support for the software can be obtained by contacting:

Stat-Ease, Inc.
2021 East Hennepin Ave, Suite 191
Minneapolis, MN 55413
Telephone: 612-378-9449, Fax: 612-378-2152
E-mail: support@statease.com, Web-site: www.statease.com

Updates for the Design-Ease® software packaged with *DOE Simplified* will be made available at the web page: **www.statease.com/simplified**

Design-Ease® Software

Version 6.0 for Windows

EDUCATIONAL VERSION

Distributed by

Productivity, Inc.
PORTLAND, OREGON

Introduction

Welcome to the Educational Version of Design-Ease® software from Stat-Ease, Inc. We are pleased to provide computing power for case studies and courses on design of experiments (DOE). Instructors and students like the software because it's so easy to use. The Educational Version contains all functions and features of the commercial version of Design-Ease, but use is limited to 180 days from date of installation. After installing the software, we recommend viewing the tutorials using Adobe's Acrobat Reader. Design-Ease also offers extensive on-line help.

The CD-ROM includes the following materials:

1   Design-Ease installation files
1   Tutorials in Adobe PDF format
2   Adobe Acrobat Reader shareware

Hardware Requirements

Component	Minimum Required	Recommended
Processor	Intel 486 or equivalent	Pentium 133 or higher
Media Drive	CD-ROM	CD-ROM
Hard Drive	20 MB free space	100 MB free space
RAM	16 MB Windows 95 32 MB Windows 98 32 MB Windows NT 4 48 MB Windows 2000	32+ MB Windows 95 64+ MB Windows 98 64+ MB Windows NT 4 96+ MB Windows 2000
Pointing device	Mouse	Mouse
Monitor	VGA (640 × 480)	Super-VGA (800 × 600 or higher resolution)
Operating system	Windows 95 or higher	Windows 95 or higher

Installation

1. Insert the CD-ROM into the drive. The installation should automatically start via the auto-run feature of Windows. If it doesn't start automatically, then

    1.1. Open the **My Computer** folder by double clicking the icon on the desktop.

    1.2. Open the CD-ROM folder in the same manner.

    1.3. Launch the **Install.exe** program by double clicking on its icon.

2. The Stat-Ease logo appears, followed by the main menu, which offers three choices. Click **Install Design-Ease 6 P**. Click **OK** to install the software.

3. The installation program loads and displays a welcome message. Click **Next** to continue.

4. By default, the software is installed on **C:\Program Files\DE6P**. If you want a different folder, use the Browse option to select your desired location. Click **Next** to continue.

5. The installation program asks you to select a program group. The default adds a Stat-Ease group to your **Start** menu under the **Programs** submenu. Design-Ease is in that group. A "Design-Ease 6 P" menu item will be added to the Add/Remove Programs list in case you want to remove the Design-Ease software from your computer.

6. The program is now ready to be installed. Press **Next** to begin the copying process.

7. Click **Finish** to return to the Stat-Ease main menu.

8. If you do not already have this software, click **Install Adobe Acrobat Reader 4.0**. (You will need it to view the Design-Ease tutorials.)

9. Select **Exit** when finished.

Using the Tutorials (recommended)

1. Double click on the **Design-Ease User Guide** icon in the Stat-Ease group. Adobe Acrobat Reader is launched automatically.

   – Or –

   Click on the Adobe Acrobat Reader software icon; next click on **File**, **Open** and select the desired tutorial from the **DE6P/User Guide** folder. The file names correspond to sections in the user manual. For example: DE02-Factor-One.pdf refers to section 2 on one-factor experimental designs.

2. You may find it easier to print the contents so it can be referred to while operating the Design-Ease program. In any case, you need the reader program to view or print the tutorials.

Technical Support

For technical support, please contact Stat-Ease at 612-378-9449 or support@statease.com.

BOOKS FROM PRODUCTIVITY, INC.

Productivity, Inc. publishes and distributes materials on continuous improvement in productivity, quality, and the creative involvement of all employees. Many of our products are direct source materials from Japan that have been translated into English for the first time and are available exclusively from Productivity, Inc. Supplemental products and services include membership groups, conferences, seminars, in-house training and consulting, audio-visual training programs, and industrial study missions. Call toll-free 1-800-394-6868 for our free catalog.

5 PILLARS OF THE VISUAL WORKPLACE
The Sourcebook for 5S Implementation
Hiroyuki Hirano

In this important sourcebook, JIT expert Hiroyuki Hirano provides the most vital information available on the visual workplace. He describes the 5S's: in Japanese they are *seiri*, *seiton*, *seiso*, *seiketsu*, and *shitsuke* (which translate as organization, orderliness, cleanliness, standardized cleanup, and discipline). Hirano discusses how the 5S theory fosters efficiency, maintenance, and continuous improvement in all areas of the company, from the plant floor to the sales office. This book includes case materials, graphic illustrations, and photographs.
ISBN 1-56327-047-1 / 377 pages, illustrated / $85.00 / Order FIVE-B0001

BECOMING LEAN
Inside Stories of U.S. Manufacturers
Jeffrey Liker

Most other books on lean management focus on technical methods and offer a picture of what a lean system should look like. Some provide snapshots of before and after. This is the first book to provide technical descriptions of successful solutions and performance improvements. The first book to include powerful first-hand accounts of the complete process of change, its impact on the entire organization, and the rewards and benefits of becoming lean. At the heart of this book you will find the stories of American manufacturers who have successfully implemented lean methods. Authors offer personalized accounts of their organization's lean transformation, including struggles and successes, frustrations and surprises. Now you have a unique opportunity to go inside their implementation process to see what worked, what didn't, and why. Many of these executives and managers who led the charge to becoming lean in their organizations tell their stories here for the first time!
ISBN 1-56327-173-7/ 350 pages / $35.00 / Order LEAN-B0001

Productivity Press   www.productivitypress.com   1-800-394-6868

BUILDING ORGANIZATIONAL FITNESS
Management Methodology for Transformation and Strategic Advantage
Ryuji Fukuda

The most urgent task for companies today is to take a hard look at the future. To remain competitive, management must nurture a strong capability for self-development and a strong corporate culture, both of which form part of the foundation for improvement. But simply understanding management techniques doesn't mean you know how to use them. You need the tools and technologies for implementation. In *Building Organizational Fitness*, Fukuda extends the power of his managerial engineering methodology into the context of the top management strategic planning role.
ISBN 1-56327-144-3 / 250 pages / $65.00 / Order BFIT-B0001

CORPORATE DIAGNOSIS
Setting the Global Standard for Excellence
Thomas L. Jackson with Constance E. Dyer

All too often, strategic planning neglects an essential first step and final step-diagnosis of the organization's current state. What's required is a systematic review of the critical factors in organizational learning and growth, factors that require monitoring, measurement, and management to ensure that your company competes successfully. This executive workbook provides a step-by-step method for diagnosing an organization's strategic health and measuring its overall competitiveness against world class standards. With checklists, charts, and detailed explanations, *Corporate Diagnosis* is a practical instruction manual. Detailed diagnostic questions in each area are provided as guidelines for developing your own self-assessment survey.
ISBN 1-56327-086-2 / 115 pages / $65.00 / Order CDIAG-B0001

HOSHIN KANRI
Policy Deployment for Successful TQM
Yoji Akao (ed.)

Hoshin kanri, the Japanese term for policy deployment, is an approach to strategic planning and quality improvement that has become a pillar of Total Quality Management (TQM) for a growing number of U.S. firms. This book compiles examples of policy deployment that demonstrates how company vision is converted into individual responsibility. It includes practical guidelines, 150 charts and diagrams, and five case studies that illustrate the procedures of *hoshin kanri*. The six steps to advanced process planning are reviewed and include a five-year vision, one-year plan, deployment to departments, execution, monthly audit, and annual audit.
ISBN 0-915299-57-7 / 241 pages / $65.00 / Order HOSHIN-B0001

LEARNING ORGANIZATIONS
Developing Cultures for Tomorrow's Workplace
Sarita Chawla and John Renesch, Editors

The ability to learn faster than your competition may be the only sustainable competitive advantage! A learning organization is one where people continually expand their capacity to create results they truly desire, where new and expansive patterns of thinking are nurtured, where collective aspiration is set free, and where people are continually learning how to learn together. This compilation of 34 powerful essays, written by recognized experts worldwide, is rich in concept and theory as well as application and example. An inspiring follow-up to Peter Senge's groundbreaking bestseller *The Fifth Discipline*.
ISBN 1-56327-110-9 / 571 pages / $35.00 / Order LEARN-B0001

A NEW AMERICAN TQM
Four Practical Revolutions in Management
Shoji Shiba, Alan Graham, and David Walden

For TQM to succeed in America, you need to create an American-style "learning organization" with the full commitment and understanding of senior managers and executives. Written expressly for this audience, *A New American TQM* offers a comprehensive and detailed explanation of TQM and how to implement it, based on courses taught at MIT's Sloan School of Management and the Center for Quality Management, a consortium of American companies. Full of case studies and amply illustrated, the book examines major quality tools and how they are being used by the most progressive American companies today.
ISBN 1-56327-032-3 / 598 pages / $50.00 / Order NATQM-B0001

PERFORMANCE MEASUREMENT FOR WORLD CLASS MANUFACTURING
A Model for American Companies
Brian H. Maskell

If your company is adopting world class manufacturing techniques, you'll need new methods of performance measurement to control production variables. In practical terms, this book describes the new methods of performance measurement and how they are used in a changing environment. For manufacturing managers as well as cost accountants, it provides a theoretical foundation of these innovative methods supported by extensive practical examples. The book specifically addresses performance measures for delivery, process time, production flexibility, quality, and finance.
ISBN 0-915299-99-2 / 429 pages / $55.00 / Order PERFM-B0001

QUICK RESPONSE MANUFACTURING
A Companywide Approach to Reducing Lead Times
Rajan Suri

Quick Response Manufacturing (QRM) is an expansion of time-based competition (TBC) strategies, which use speed for a competitive advantage. Essentially, QRM stems from a single principle: to reduce lead times. But unlike other time-based competition strategies, QRM is an approach for the entire organization, from the front desk to the shop floor, from purchasing to sales. In order to truly succeed with speed-based competition, you must adopt the approach *throughout* the organization.
ISBN 1-56327-201-6/ 560 pages / $50.00 / Order QRM-B0001

SUPPLY CHAIN DEVELOPMENT FOR THE LEAN ENTERPRISE
Interorganizational Cost Management
Robin Cooper and Regine Slagmulder

It is no longer sufficient for a lean enterprise to be highly effective. In today's competitive business climate, a lean enterprise must also be part of a highly efficient supply network. Interorganizational cost management is a structured approach to coordinating the activities of firms in a supplier network to reduce the total costs in the network. In this book, Cooper and Slagmulder use nine case studies to document how successful companies transfer cost management pressures across organizational boundaries. Suppliers are also a major source of innovation for lean enterprises. Successful supplier networks encourage every firm in the network to innovate and compete more aggressively. Read this book to learn how to manage your supply chain to forge a competitive advantage while reducing costs.
ISBN 1-56327-218-0 / 544 pages / $50.00 / Order COSTB2-B0001

TARGET COSTING AND VALUE ENGINEERING
Robin Cooper and Regine Slagmulder

What would happen if everyone in your company followed a disciplined approach to cost reduction? How can it be done? With smart cost management. Two powerful strategies—target costing and value engineering—will get you well on your way. You will find both in this comprehensive book, the first in a series guaranteed to increase your profits. Effective cost management must start at the design stage. As much as 90–95% of a product's costs are designed in, meaning they cannot be avoided without re-designing. That is why effective cost management programs focus on design and manufacturing. The primary cost management method to control cost during design is a combination of target costing and value engineering.
ISBN 1-56327-172-9 / 400 pages / $50.00 / Order COSTB1-B0001

Productivity, Inc. Consulting and Public Events

EDUCATION...IMPLEMENTATION...RESULTS

Productivity, Inc. is the leading American consulting, training, and publishing company focusing on delivering improvement technology to the global manufacturing industry.

Productivity, Inc. prides itself on delivering today's leading performance improvement tools and methodologies to enhance rapid, ongoing, measurable results. Whether you need assistance with long-term planning or focused, results-driven training, Productivity, Inc.'s world-class consultants can enhance your pursuit of competitive advantage. In concert with your management team, Productivity, Inc. will focus on implementing the principles of Value-Adding Management, Total Quality Management, Just-in-Time, and Total Productive Maintenance. Each approach is supported by Productivity, Inc.'s wide array of team-based tools: Standardization, One-Piece Flow, Hoshin Planning, Quick Changeover, Mistake-Proofing, Kanban, Problem Solving with CEDAC, Visual Workplace, Visual Office, Autonomous Maintenance, Overall Equipment Effectiveness, Design of Experiments, Quality Function Deployment, Ergonomics, and more! And, based on continuing research, Productivity, Inc. expands its offering every year.

Productivity, Inc.'s conferences provide an excellent opportunity to interact with the best of the best. Each year our national conferences bring together the leading practitioners of world-class, high-performance strategies. Our workshops, forums, plant tours, and master series are scheduled throughout the U.S. to provide the opportunity for continuous improvement in key areas of lean management and production.

Productivity, Inc. is known for significant improvement on the shop floor and the bottom line. Through years of repeat business, an expanding and loyal client base continues to recommend Productivity, Inc. to their colleagues. Contact Productivity, Inc. to learn how we can tailor our services to fit your needs.